DIVIDE

DIVIDE

The Relationship Crisis
Between Town & Country

ANNA JONES

K

For Avril

An Hachette UK Company
www.hachette.co.uk

First published in Great Britain in 2022 by
Kyle Books, an imprint of Octopus Publishing Group Limited
Carmelite House
50 Victoria Embankment
London EC4Y 0DZ
www.kylebooks.co.uk
www.octopusbooksusa.com

ISBN: 9780857839732

First published in paperback in 2023

Distributed in the US by Hachette Book Group, 1290 Avenue of the Americas,
4th and 5th Floors, New York, NY 10104

Distributed in Canada by Canadian Manda Group, 664 Annette St.,
Toronto, Ontario, Canada M6S 2C8

Publisher: Joanna Copestick
Publishing Director: Judith Hannam
Editor: Samhita Foria
Design: Paul Palmer-Edwards
Production: Nic Jones and Lucy Carter

A Cataloguing in Publication record for this title is available from the British Library

Printed and bound in Great Britain

10 9 8 7 6 5 4 3 2 1

PROLOGUE

I

PROLOGUE

As a child growing up in the 1990s, I dreamed of leaving the countryside and moving to New York City. In fact, I became quite obsessed with the idea.

I'd watched a movie on TV about a high-flying career woman, played by Diane Keaton, who strides around Manhattan in a pencil skirt and heels. I got so swept up in the fantasy that I'd wear Mum's stilettos, stuffed with socks to stop them falling off my feet, and march around the garden with her handbag slung over my shoulder. I couldn't quite conjure the thrill of Madison Avenue on the lawn, so I'd pick my way across the field, past the tractor, feed troughs and our blind old sheepdog Jaff's kennel and stomp around outside the cowshed. The shoes made a better clip-clopping sound on the concrete by the muck heap.

I fantasised about city life throughout my teenage years. At the age of 18, I left our small beef and sheep farm on the border between England and Wales to study journalism at the University of Central Lancashire. I remember walking down Fishergate in Preston feeling like I'd made it to Manhattan, just like Diane Keaton. It didn't matter to me that it was a town in the north of England; all I wanted were people, crowds, traffic, bustle, and noise. I felt a thrill at the sound of sirens. This was the urban life I'd dreamed of.

The great irony, which has only just occurred to me, is that Diane Keaton's character in the 1987 romantic comedy *Baby Boom* ends up leaving New York and moving to a farmhouse in Vermont. By my early thirties, the novelty of city living had well and truly worn off for me too. I'd moved to Birmingham, then Manchester and Bristol, building my career at the BBC, though always avoiding London as one step too deep into the urban jungle, and by the time I'd racked up 21 different addresses, something deep inside was calling me home. Back to my roots on the Welsh Borders; back to the farm.

Sadly, real life is more complicated than shoving your belongings in an old Volvo and moving to Vermont. The road back to rural life is fraught with obstacles – what would I do for work? Could I commute? Where would I live? And most profoundly of all – would I fit in? I'm not the person I once was. I am urbanised – used to convenience, used to choosing from endless options, used to living and thinking a certain way. I have a life that is different in almost every way to my conservative, rural, working-class upbringing. I am no longer part of the community I grew up in. Could I ever, truly, belong again? Can a townie ever go back to being a country dweller?

Only when I started pondering these questions did the barriers between my urban and rural life, and the conflict between my urban and rural self, reveal themselves. They were subtle and opaque – difficult to grasp and articulate and so easy to overlook and ignore. But they are real. And the more I looked, the sadder I felt. I'd uncovered a hidden, chasmic divide. And once I'd seen it, I couldn't unsee it.

CHAPTER ONE

HOME

PHYSICAL WORK, toil and hard graft on the land have been the mainstay of life in my family for hundreds of years. Farm work has turned generation after generation of strong and strapping young Joneses into stooped and cash-strapped old men and women. We're better at inheriting arthritis than money; but with their aching joints and modest bank accounts so too came pride, identity, belonging, community and, perhaps most importantly of all, stories. These are the ties that have bound my family to the hills and villages, fields and farms along one small section of the English/Welsh border since the 1700s, and probably long before.

My family, on both sides, have always farmed. I found my great-great-great-grandfathers in the 1861 census, farming just a few miles from each other. On my mother's side, John Roberts was living with his wife Anne, three of his six children and his elderly pauper father, Richard Roberts, at Hirnant in the historic county of Montgomeryshire, now Powys, in Mid Wales.

One bitterly cold January day, I went off in search of his farm.

Hirnant is a sleepy hamlet snuggled away at the foot of a gently sloping valley which, like most Welsh hills, is either grazed by sheep or carpeted in conifer trees. A clear, rocky stream, the Afon Hirnant, babbles coquettishly past gloriously restored white-washed nineteenth-century cottages. They're begging to be plastered across estate agent brochures, sparking envious daydreams of escaping to the country, and prompting nosey parker poke-abouts into neighbouring property prices.

My great-great-great-grandfather's modest little farmworker's cottage is a now a beautiful smallholding and equestrian property with fenced pastures, a vineyard, extensive gardens and a waterfall. The price of my ancestor's old home is way out of reach for most local people, yet, ironically in the mid-1800s, it would have provided the most basic and meagre accommodation to John Roberts and his family of nine. In 1861, he was a 50-year-old tenant farmer of six acres, having gone up in the world slightly since the 1851 census where he is listed as an agricultural labourer – the same low-paid profession that had left his father, born in 1777, and goodness knows how many generations before that, destitute. John Roberts' descendants, in my mother's direct family line, remained farm workers for generations; their lives teetering precariously on the poverty line throughout the nineteenth and early twentieth centuries.

Fortune smiled a little brighter on my father's side of the family. In 1861, while John Roberts scratched a living high up in the hills, my paternal great-great-great-grandfather, William Jones, farmed on lower ground, 12 miles east along the lush and fertile Tanat Valley and just a hair's breadth over the border in the Shropshire parish of Llanyblodwel, which, after decades of skirmishes, settled in the hands of the English in 1536. William was a dairy farmer, a deeply committed Wesleyan Methodist and a proud Welsh patriot who lamented the increasing dominance of the English language in his borderland community. In 1852 he funded the extension of an entirely Welsh-speaking chapel at Cefn y Blodwel, where a simple marble memorial (though still on the grander side of what's usual in spartan Welsh graveyards) bears his name to this day.

In 1861, William is listed as a tenant farmer of 132 acres, employing '13 men'. He led a virtuous life with his wife Margaret, seven children and six employees – a farm labourer, cowman, dairy maid, house servant, and two carters. William's home is still a working farm, though in the hands of a different local family. The Old Cart House, where William's team of carters looked after the wagons and horses, is now a beautifully renovated holiday cottage.

It was a nice surprise to learn that William Jones was a bit of a local celeb, known for his philanthropic endeavours, agricultural prowess and generally being an all-round good egg who was, literally translated, the 'embodiment of kindness'. In 1906, the Methodist minister and author, Owen Madoc Roberts, wrote a book in Welsh, *Tri Brawd*, meaning 'Three Brothers', about William Jones and his siblings John and Richard. According to his admiring biographer, William had an instinct for farming: 'Mr Jones worked hard and dug drains, laid pipes to drain the land. He straightened hedges and flattened fields. He took pleasure in feeding his animals and looking after his machinery and consequently was respected as one of the best and most revered farmers in the area.'

These days, that kind of land use change could get old William in trouble with Natural England. Draining wetland and messing about with hedges – he'd certainly wind up on the environmental naughty list. But in 1858, such was his popularity, the community rallied round to help when a mystery disease claimed 17 of William's milking cows. Not to be beaten, he turned to sheep farming instead – a career continued by his descendants, including my dad, to this day.

William Jones was obviously better off than John Roberts, but I like to think my great-great-great-grandfathers might have known each other; perhaps buying and selling livestock at the markets in Oswestry or Welshpool, passing each other on the roads in and out of town or nodding hello at an agricultural show, summer fête or religious festival.

The poorer Roberts family persevered with agricultural labouring and subsistence farming down the generations. In 1908, John's granddaughter, Hannah Roberts, married his grandson, Thomas Jones (my maternal great-grandparents were first cousins – one of the more uncomfortable discoveries made while researching this book).

Thomas and Hannah raised 12 children in a small farmworker's cottage at Wern Ddu near the village of Llansilin in Powys, just a mile or so down the road from where my parents still farm today. Thomas was a farm labourer and a molecatcher, while Hannah – who was continually pregnant or breastfeeding for 17 years – earned what she could selling eggs and home-churned butter in Oswestry, a market town five miles away. She was, by all accounts, a natural storyteller with a wicked sense of humour and a flair for writing. Stories have passed down our family of Hannah bumping along in her horse and cart composing poems, plays and songs and rehearsing her brood of musical children, like the Welsh von Trapps, for their next Eisteddfod appearance – cultural festivals of music, dance and poetry with prizes awarded for the best recitals and performances. Thomas and Hannah's seventh son was my grandfather, Wilfred. Perhaps put off by the poverty and hardship he'd known as a child, he was the first in my mother's direct line to break from farming and became a butcher instead.

My father's side of the family, in contrast, stuck with agriculture and even managed to transition from tenants to owning a small farm of 65 acres called 'Craigllwyn'. It was bought for my great-grandfather John Wddyn Jones, the grandson of William Jones – that revered farmer who embodied kindness.

John Wddyn, a small and mild-mannered man, was born with clubfoot and so walked with a limp. I was always curious about him as a child and whenever I asked questions he was described – in terms no longer appropriate – as 'weak' or 'a bit crippled', but also 'kind' and 'well-liked'. He moved to Craigllwyn in the early 1920s with his fiery and domineering younger wife Elsie (my great-grandmother, 'Nanny', who plays an important role in our family farming story) and his young family, including two-year-old Bill (my grandfather), a gentle, softly spoken boy who took after his father in temperament.

Craigllwyn is a beautiful upland farm, 'on the top' as locals say, with views over Llansilin towards the Berwyn Mountains. It has good outbuildings, patchwork fields of permanent pasture bordered with thick hedgerows and a large nineteenth-century stone farmhouse, though it came without the carters and house servants. The Agricultural Depression of the 1930s hit the family hard and Craigllwyn Farm struggled thereafter. Nanny and Grandad Bill, who farmed in partnership following the death of John Wddyn in 1950, ran a small dairy herd of 18 British Friesian cows, a flock of 100 native Welsh ewes, 12 native Welsh sows, one Welsh boar, chickens, ducks and a dozen turkeys. Still, the farm alone could not support them, especially since Grandad Bill had married in 1951 and started a family of his own. To make ends meet, he got a job in the neighbouring limestone quarry, tipping stone into the crushers. He milked his cows at dawn, walked to the quarry, worked his shift, and trekked home at dusk to milk the cows again. My grandmother, Beryl, looked after the farmhouse, raised eight children (the eldest being my father) and worked night shifts as a nurse at the hospital in Oswestry, our nearest town.

By the 1920s, two branches of my family tree were living and working just a mile or so apart. My grandfathers – Wilfred at Wern Ddu and Bill at Craigllwyn Farm – were schoolmates at Ysgol Bro Cynllaith, the village primary school in Llansilin, and played together as boys. Little did they know, these two farmer's sons both named Jones, that Wilfred's youngest daughter, Avril, would one day marry Bill's eldest son, Tony – my parents.

Our family story is far from unique. This deep-rooted interconnectedness of families in rural communities is a familiar tale right across the British Isles. It's a heritage shared by thousands; common enough to give rise to some of the crueller stereotypes about country people and jokes about inbreeding and backwardness. 'Welsh people liked to keep their farms in the family,' Mum winks and taps her nose whenever I ask about the marriage of cousins in our own family tree.

People can be hefted, just as much as the sheep that graze the Welsh mountains or Cumbrian fells. My ancestors stayed, perhaps against their best interests, to eke a living as tenants and labourers on land that did not belong to them or on small farms that returned little or no profit, and watched their children and grandchildren do the same.

I RECOGNISE A SHARED 'hefting' experience immediately when I met Ruth Grice, a dairy farmer from Melton Mowbray in the Leicestershire Wolds. Like me and my two younger sisters, she is one of three girls born and raised on a small farm that belonged to her grandfather and great-grandfather. They bought it in the 1950s, but her family, just like mine, has been rooted in the local area working in agriculture for generations, moving only between three neighbouring villages since the late 1700s. Ruth also left rural Leicestershire as soon as she had the chance, chasing urban adventures at university in Sheffield and later settling in Oxford. For 10 years she never looked back, immersed in a successful marketing career and a busy city lifestyle. But around the age of 30, exactly the same time I felt it, a gut feeling started to niggle and, according to Ruth, 'a light switch came on.'

'I don't even know what made me look at a prospectus,' she says, 'but I ended up going to an open day at the Royal Agricultural University in Cirencester.'

Ruth was accepted onto a Master of Business Administration (MBA) course in Advanced Farm Management, 'absolutely loved it', then returned to the family farm to work alongside her father and 180 pedigree Holstein cows.

'I just felt this gut instinct that it was the right thing to do,' she says. 'It's the tie to the land and the animals and . . . I can't describe it.'

I urge her to try and put it into words, hoping she can help me understand this mysterious pull towards home, family and roots.

'Ok, this is going to sound really weird . . . I remember watching an episode of *Our Yorkshire Farm* and she described the piece of land that she and her family farmed as their "heaf" – although she pronounced it like "hoof". It's a phrase I'd never even heard of, maybe it's a Yorkshire thing – but that's the boundary of the land you know and cultivate. And that's how I feel about here. When I'm driving down Mum and Dad's lane, just for a little way, that's our farm's land. And I get this sense of, "Oh, I'm home." I'm in the mothership. It's my happy place.'

'I think it's love' she adds thoughtfully. 'Love for what my parents have done for me and my sisters, love for what they have done for the land, for the farm and just a desire to carry that on'.

I recognise what Ruth describes, I've felt it myself, but, if you're lucky enough to have had a happy childhood home, irrespective of whether you grew up on a farm or in the city, surely, it's the same for everyone? Feeling a deep emotional connection to a place, to your 'heaf', is not the preserve of multi-generational family farms, nor is it exclusively a rural experience.

There's a chap called Alan who we know from our allotment in Bristol, not too far from the city centre. He's in his seventies and has tended the same vegetable patch since he helped his grandfather as a 10-year-old boy. A patch of urban earth has passed down the generations, just like the farms in our countryside. Alan makes the same quip about dropping dead and getting buried in the raised beds every time we see him.

That magnetic pull of home ground is a powerful force for millions. Our next-door neighbour, another Alan, is a fourth-generation Bristolian, the son of a crane driver who worked at the Bristol Aeroplane Company during World War Two, 'moving bits of military aircraft around'. Alan is hefted to the streets of his city. He was born and raised in a house just off Jamaica Street in Stokes Croft in 1944. He moved to the suburbs for a few years before returning to the inner-city in 1980, where he's remained ever since. He had jobs in London and Reading but remained firmly rooted in Bristol, despite years of exhausting commutes. When I ask him why he didn't move, he shrugs: 'History and continuity I suppose'.

It's exactly the same language I hear on family farms up and down the country, but it's not something Alan consciously thinks about. He seems slightly bemused by my questions and wouldn't in a million years describe Bristol as his 'heaf'. It leads me to wonder if the culture I've been brought up in, of frequently and emotionally verbalising our connection to the land, is a uniquely rural experience. Maybe it's the narrative, not the land, that steeps us in a place? The stories we tell ourselves. I need to ask the Yorkshire Shepherdess herself about this heaf thing.

Amanda Owen is star of the hit Channel 5 series *Our Yorkshire Farm* and has written several books about her life at Ravenseat, a remote hill farm in Swaledale, with her husband Clive and nine children. In the dark depths of the winter 2021 lockdown, Amanda is giving a live talk and Q&A on YouTube to help raise funds for the rural charity

Farming Community Network. On my behalf, Adam Bedford from the National Farmers' Union (NFU) asks my question about this intriguing idea of a 'heaf'.

'I think it's a Yorkshire term,' she replies. 'It's that sense of belonging to land.'

I know the Cumbrians say 'heft' and, in Wales, there are two beautiful words: 'cynefin' meaning 'habitat' or being connected to a place; and 'hiraeth', which describes a sad or nostalgic longing for home. 'The sheep know their heaf,' she continues, 'and they know where they are raised, and they know the boundary keeping them there. That's an ancient practice that has gone on forever. In a way, maybe a person can be heafed?'

And then Amanda says it – the phrase I and probably every other farm kid in the world has been raised on: 'Farming is a way of life.'

There is a subtle language shared by farming folk the length and breadth of the British Isles – a kind of poetic stoicism, bursting with pride, a whiff of nostalgia, and a hint of martyrdom. I recognise it instantly every time I hear it and my rural soul soars to the sound of it. It takes me home. My urban soul cringes slightly. It can get a bit 'naval-gazey' at times.

'Maybe that sounds a bit clichéd,' says Amanda, 'but it goes a lot further than that – to the children in the local schools, to the jobs in the area, to the tourism. They are all linked by farming.'

My ears prick up when she ponders whether she, herself, could claim to be 'heafed'. She was born in Huddersfield and doesn't come from a farming family – but she sees that separation as a benefit:

'I've seen different things, lived a different life and I know what to appreciate.'

I've had countless conversations over the years about whether a farming background makes you a better farmer. Personally, I don't think it matters if you've been heafed or hefted to a patch of land for generations or just a handful of harvests – it's how you look after it that matters. Multi-generational land management doesn't always translate into a deep spiritual connection with the land or, indeed, good farming. Sometimes it can even breed discontent and laziness. One of the greatest criticisms of being born into anything is that you risk taking it for granted. I didn't appreciate the farm when I was a teenager – I was total crap. I had no interest in checking on the in-lamb ewes or putting fresh water in their pens. I wanted to watch Ant and Dec on *SMTV Live*. If I had been made to stay at home and work with Dad, I would have grown into a bitter, negligent, and resentful farmer – even if I could boast I was the sixth or seventh generation. Now, after pursuing my own dreams and becoming my own person, I'm always asking Dad for a job. I had to leave it to love it.

I'm not the only one who's had a boomerang relationship with the family farm. Twenty-five-year-old Gethin Bickerton would have been the next generation if he'd decided, like his older brother, to follow in his parents' footsteps. Their farm is only five miles away from my family home and our childhoods, despite the 14-year age gap, followed pretty much the same trajectory – we both grew up on traditional beef and sheep farms, went to the same rural Welsh-speaking high school in Llanfyllin and joined the same Young Farmers Club (Dyffryn Tanat). I like to think we knew how lucky we were – but did we?

'If you go up to one of the fields on the mountain there are 360-degree views of pure countryside, it's absolutely gorgeous,' says Gethin, picturing his home over a Zoom call from his flat in Cardiff. 'But I'll happily admit that I didn't appreciate it when I was younger. Primary school and high school is the time when young people need to make connections with other kids, but I couldn't do that in the same way because I lived so rurally. I couldn't "pop round" to see friends or "pop into town" – you can't "pop" anywhere! I had to rely on lifts, so it took me a while longer to develop my independence.'

Gethin felt hemmed in by the repetitive cycle of farming life, though he prefers to call it a 'strong routine', which sounds less negative. From a young age he yearned for something different. He bolted straight after his A-Levels and headed for Cardiff, home to nearly a quarter of the Welsh population, to study theatre and drama. He became an actor, got work straight after graduating and now travels all over Wales performing. He stars in a one-man bilingual show *Llywelyn Ein Llyw Olaf* playing Llewellyn the last Prince of Wales, and also toured Welsh primary schools teaching children about food hygiene and safety:

'It was based on a spaceship; I was Captain Nebula, and I had a helmet – what more could you want?'

Landing a dream role in a West End musical would probably top Captain Nebula. In 2019, Gethin signed up with a London-based acting agent and his career looked set to boom. Nothing could stand in his way ... except maybe a global pandemic. In March 2020 theatres closed and work dried up completely for actors like Gethin. He escaped to the family farm in late February, bracing himself for lockdown, and ended up staying there for seven months.

'The thought of being stuck in a flat with two other people and no garden – I dread to think how my mental health would have been impacted,' he says. 'For that I will be forever thankful for where I grew up. I honestly couldn't have been happier.'

Lockdown fundamentally changed Gethin's relationship with home – he renovated old buildings, built a water feature in the garden (something his Mum had wanted for 20 years) and helped his Dad and brother with the livestock every day. He started to see the farm, and his community, through fresh eyes and felt his own perceptions changing:

'It's really easy and quite wrong of us to think that country life lags behind city life. I'd always thought the countryside is about 10 years behind the times, but it was a real eye opener and a beautiful experience to see that it isn't. My local Young Farmers' Club is doing amazing work on mental health and, in terms of my own sexuality, it was great hearing that there are people who are able to be themselves and their sexuality doesn't need to be something to hide.'

It wasn't what he had expected and Gethin warmed to the subtle and surprising changes that had taken place while he'd been away:

'I did a workshop at my old high school and heard that there were now openly gay teachers and transgender pupils. I became emotional speaking with my drama teacher, who was the first person I came out to when I was 17. The sense of relief I felt was overwhelming, to hear that coming out and being your true self is now deemed more acceptable than it was back then. Of course, living in Cardiff, it's all fine because everyone's more aware of these issues, but it was so nice to see it is happening at home too. That pride I feel in Cardiff, I can now feel back home too.'

The months ticked by and Gethin allowed himself to be swept along with the farming calendar, through lambing, calving, silaging, and hay making; a routine that had once felt so stifling, like a trap, now brought him a new sense of freedom and discovery. By the time September arrived, and it was time to return to Cardiff, Gethin didn't want to leave.

'I really missed home; in a way I'd never missed it before. I thought, how can I be here in Cardiff and still be connected to home? How can I help people without actually being there?'

The answer came when he was driving down the A470 and saw some black plastic-wrapped silage bales bearing the message 'Share the Load.' It's a 24/7 confidential counselling helpline run by the farming mental health charity, The DPJ Foundation, which supports hundreds of farmers across Wales struggling with poor mental health. The charity was set up in 2016 following the death of Pembrokeshire agricultural contractor Daniel Picton-Jones, who took his own life after a long battle with anxiety and depression. Gethin now sits on the board of trustees, using his superfast, inner-city broadband to attend virtual meetings, enabling him to plug straight back into his rural community.

'It's definitely thanks to the pandemic and spending all that time at home,' he says. 'The amount of people that call the helpline is great. I didn't have anything like this when I was growing up. I wonder if I'd had somebody, externally, to talk to, if that would have helped me in some way?'

Rural life can be lonely and isolating, and the greatest privilege is being able to leave it, find yourself and return home again to a loving and welcoming family. Gethin and I know how lucky we are – it's a winning lottery ticket in life – but it's not the same for everyone.

For the many thousands who choose to stay on the family farm, and thrive, there are always those who get stuck – imprisoned by a lack of education, confidence, encouragement, or opportunity; trapped by their own sense of duty to stay behind. I've seen it many times and it makes me sad, these lives sacrificed to the farming 'way of life'. We should be careful of overselling it, with this poetic language of rural romanticism. That path was laid out for me, if I'd wanted it, but it would have been a mistake to follow it. Wrong for me and most definitely wrong for our farm.

Confusing farming heritage with farming talent is a dangerous pitfall; so too is the all-too-common assumption that no agricultural background means no agricultural ability.

There are new entrants, first-generation farmers and 'incomers' to the countryside who feel just as tied to the land as their 'native' neighbours, mainly because it took them so long to find it and they worked so hard to get it.

SINEAD FENTON GREW UP in Newham, East London. Her only connection to farming was the sight of the arable fields she saw from the train window on day trips to Southend-on-Sea:

'I didn't understand what they were. They were just square patches in the ground that you didn't really question. That was the extent of my engagement with the food system. I didn't care about food – it was just a necessity.'

Sinead never knew her father and her mum, a school nurse in a primary school, raised her alone on a low income. Meals, though homecooked, were about survival, not enjoyment.

When Sinead went to university in Leicester she watched in fascination as her housemates found pleasure in cooking and eating food. She discovered the simple joy of a homemade curry made from fresh ingredients, herbs, and spices. On her friend's dad's allotment, she saw a vegetable growing in the ground for the first time. Asparagus transformed, before her eyes, from an inanimate plastic-wrapped supermarket item into a wondrous living thing. It fed her fascination with 'all things that come out of the earth', leading to a geology degree and a disappointing career in the mining industry. Disillusioned, she took voluntary redundancy and got what she regarded as another 'shit' job in software recruitment in Hertfordshire. Still feeling lost and unfulfilled at work, Sinead would take her sandwiches and sneak into lunchtime soil seminars at Rothamsted Research, the agricultural scientific research institute just around the corner from her office. Sitting quietly at the back, hoping no one would spot she wasn't a student, something finally clicked. She'd been on the right path after all – she was meant to work in the earth – but growing food, not mining rocks. Sinead's lunchtime moonlighting sparked a new career epiphany – and this time it would be third time lucky – she was meant to be a farmer. She took voluntary redundancy a second time:

'I moaned so much they paid me to leave!' she laughs.

Sinead went to work on a struggling community garden and allotment site in Epping, agreeing to run it on a voluntary basis with her partner Adam. She supplemented two-and-a-half years of unpaid work with various part-time jobs, including 14 hour shifts as a pot washer at a cookery school.

'If you're not from a farming background but you want to get into it, that's how you do it,' she says, 'through unpaid growing work and then working excessively elsewhere to get a bit of money.'

Sinead and Adam eventually wound up the community garden to search for something more permanent, where they could make a real life for themselves. At the Oxford Real Farming Conference in 2019, they crossed paths with the Ecological Land Co-Op, an organisation which helps new entrants get a leg-up in the farming industry by offering long-term leases on small plots of land:

'We didn't think we'd be considered because we weren't from farming backgrounds and we didn't have a lot of money, but they were keen for us to become land stewards.'

Sinead and Adam developed a business plan for a market garden producing more than 70 different varieties of vegetables and edible flowers for the restaurant trade and, in 2020, they finally said hello to their forever home: a four-and-a-half-acre former arable field in East Sussex with a 150-year lease and permission for a temporary dwelling. They arrived in March, two days before the first national lockdown, faced with an empty field full of weeds, no infrastructure, and no water. They bought a freezing cold caravan and dug a trench for an internet cable:

'We got broadband before we got water, which took 11 months,' says Sinead. 'I personally would have chosen water, but I got to watch lots of Netflix and not wash.'

They're like modern-day homesteaders, stamping their claim on a patch of land and ploughing every penny and every ounce of energy into making it a success. I'm amazed at their tenacity.

Their big adventure, which sounds a bit like the uplifting plot of a quirky feelgood film, actually coincided with one of the toughest times in Sinead's life. Years of stress, overwork and rushing between jobs in London had taken a severe toll on her mental health, dumping her in a fog of depression and anxiety. Far from leaping joyfully into her exciting new life, she dragged herself over the threshold, exhausted and spent.

'When we came here the field was silent,' she says. 'There weren't many signs of life. If you dug a hole, there weren't many worms, and there weren't many birds around either. Land shouldn't be silent – it's a sign that something's wrong – whereas for me it was the complete opposite. I was desperate for silence because what was going on in my head was so noisy.'

From where I'm sitting, it sounds like a gruelling experience. Winding up in an empty field devoid of all infrastructure and even the most basic means for survival? Battling depression in a cold caravan in winter with no hot water? Hellish. But, to Sinead, the field brought salvation, stillness, and rock-solid permanence – a home:

'It's given me purpose and a reason to get up each day. I was troubled all through my twenties about not really having a place in the world. I don't know my Dad, so that's a huge part of me I know nothing about, and I struggled with my sense of identity. Being mixed-race I never felt like I belonged in different parts of society.'

Sinead has always been an outdoor girl. As a child growing up in Newham, she'd escape the noise of the city at every opportunity, exploring Epping Forest with her mum or jumping on the train to her beloved Southend-on-Sea, whizzing past those strange squares in the ground, en route to her favourite place in the world. She rejected urban life as soon as she could and knew instinctively that she'd be much happier in the countryside – but it was well-meaning friends that sowed the fear and doubt:

'The space I come from – the urban environment with lots of activists – they were the ones asking, "How do you feel going to a place that's predominantly white where you don't belong, and where you might not be accepted?" The fear came from people telling me that I should be scared, but my experience has been fine. I do belong. I felt so rootless with my identity, my race and never really felt like I had a home – but this has become that place.'

Sinead spent years searching for her rural home. Kenyan pig farmer Flavian Obiero found his almost by accident. In 2010 he got a week-long work placement on a mixed farm in Hampshire, little more than a box-ticking exercise for his university degree in Animal Management. But something clicked immediately – he enjoyed the work, particularly looking after the pigs, and the farmer offered him a permanent job after just one week. It was the beginning of Flavian's farming career and he's worked full-time in the industry ever since leaving uni.

He arrived with no preconceptions about attitudes towards race in the countryside. He'd grown up in coastal Kenya, so the idea of being singled out because of the colour of his skin was a bizarre concept to a confident 19-year-old Flavian.

'When I went to the farm, I was quite naïve to stuff like that' he says. 'But I remember walking into pubs with the farmer and everyone going quiet. People were quite standoffish and wouldn't engage with me. But once they knew me, I'd go to the pub on my own and they were fine.'

He pauses.

'There was a bit of name-calling, but I blame myself for not nipping it in the bud. Now it's something I won't entertain but back then – because I was trying to fit in and be accepted as one of the farmers – I let it go. I was called "darkie" and things like that. I should have called it out, but I didn't.'

Flavian wanted to feel at home; so much so he was willing to put up with racism. That need to belong is a powerful force – overruling even our core values.

Most of us are on, or have been on, a journey towards 'belonging somewhere', and putting down emotional roots is a beautiful part of being human. Some people are lucky enough to be born where they belong, and never have to leave, while others face a lifelong search for it. I'm still searching. I've felt torn between my urban 'home' in Bristol and my rural 'home' on the Welsh Borders for a very long time, and I'm yet to experience the epiphany described by Sinead, Ruth, Gethin and the Yorkshire Shepherdess. How I envy them.

I believe there is a fork in the road that separates the urban and rural 'belonging' experience. For instance, feeling like a 'true local', and the importance placed on being accepted as such, is, in my view, a uniquely rural experience. It is less about your own feelings of attachment to an area and more about how you are viewed within the peer system of a community.

'YOU NEED 15 GENERATIONS of your family in the local churchyard to feel like a local around here,' jokes John Yeomans, a Brummie boy turned Welsh hill farmer.

He grew up in Handsworth Wood, in the suburbs of Birmingham, the son of a butcher who went to school in Handsworth, a deprived inner-city area with high rates of unemployment, problems with overcrowding, racial tensions, and social unrest. He got used to the fights and riots that broke out regularly on Soho Road:

'I remember cycling home from school one day and the whole road was full of hundreds of youths – all these kids chucking stuff – it was absolutely terrifying. Two schools hated one school and they'd just come for a big fight. I had a racing bike with a leather saddle, and it had this big tear across it from where a stick or something had hit it. They turned over police cars and all sorts.'

Still, it was easier to avoid a massive riot outside the school gates than it was to escape the bullies who picked on him every day:

'I hated school. I was a small and podgy schoolboy who was no good at sports and not very clever.'

It was a rough start to life and John lived for Sunday afternoons, after church, when his family would go and look at farms:

'They were butchers on my dad's side really, but they had links with farming going back a generation or two and that was what he'd always wanted to do.'

John's dad, Albert, searched for years, travelling all over the Midlands looking for a place where he could farm and still commute back to his shop in Handsworth. He never gave up and finally, at the age of 60, and considerably further away than he had originally planned, the butcher from Birmingham bought 68 acres and a farmhouse near the tiny Welsh village of Adfa in Powys. In the late 1970s, he moved from a city with a population of a million people to the most sparsely populated region of Wales.

At 18, John decided to knock academia on the head and go and work with his dad on the newly acquired family farm. He enrolled at the Welsh Agricultural College (WAC) in North Wales, did a couple of farming placements and so began his transition from a shy bullied city boy into a confident and strapping young farmer. 'I grew about five inches which helped,' he smiles.

He came home to farm full-time in July 1984 'to half of our 23 cows barren and the 200 or so ewes lambing at around 90%'.

He met his wife Sarah, a Londoner, in 1986 while visiting friends in Aberystwyth. They were married a year later.

They've since expanded the farm to nearly 300 acres with 90 suckler beef cows and heifers and 700 ewes, 'with a scanning percentage up to as much as 173%, so almost double the 90% I came home to'.

John and Sarah had three sons, who all speak fluent Welsh:

'I still tell everybody I'm a townie though,' he says.

With his shaved head, oxblood red Dr Martens and a lovely lingering lilt of inner-city Birmingham, John isn't your typical Welsh mixed hill farmer. There's a worldly twinkle about him that says I've seen a few things, and he's remained a devoted fan of punk and ska all his life. During a formal interview for a prestigious farming scholarship, when asked where he sees himself in 10 years' time, John replied: 'Managing my son's band on the main stage at Glastonbury.'

In 2018, he and Sarah scooped Sheep Farmer of the Year at the *Farmers Weekly* Awards and, to date, he's the only winner to go on stage wearing Dr Martens boots and a tuxedo.

With such a mixed background I'm interested to know what John identifies as. He's not sure what he is, but he's clear about what he isn't: 'I don't think I'm Welsh for starters.'

He finds it irritating when people move into the village, learn a few phrases, and try to pass themselves off as a local. Equally, he's raised three fiercely proud Welshmen – one son is slowly plastering himself with patriotic tattoos.

What I find fascinating, thanks to his dual urban/rural passport, John has been able to straddle two divided local communities:

'In our village there's the people who have moved in and then there's the other people.'

I assume by 'others' he means 'true locals'.

'Me, Sarah and the boys, and one mate at the other end of the village, we are the only people who get invited to social events by both groups. They built a load of new houses which changed the dynamic of the village quite a lot. If you've got ten or a dozen houses and then you add another 40 – and the bulk of the people who buy them are not from around here, or even Wales, or the countryside – that changes things radically.'

Personally, I think the reason John has been so warmly accepted into the bosom of his community – transcending the invisible divide between 'incomer' and 'local' – is because he possesses two important personality traits: he's self-deprecating and he's a doer.

'I can't bear a bragger,' he says.

John doesn't blow his own trumpet, a humble quality which goes down well with many country people. My Mum can sniff out a show-off at a hundred yards and it's often confident, chatty 'outsiders', new to the area and seeking to ingratiate themselves with the locals, who unwittingly stray onto her 'Braggy Big Head' radar. She prefers to bond over trip-ups rather than triumphs. John's the same:

'I run myself down a lot, it's just the way I am. Set your sights low and prepare to be disappointed!'

John is also heavily involved with the Farmers' Union of Wales and isn't afraid to make his voice heard, doggedly banging on the doors of the Welsh Government, holding their feet to the fire on big issues like Bovine TB (a disease he's never even had in his own cattle but he feels compelled to stand up and fight for farmers who have). He's purely a grassroots member – far too humble to push himself into a top job or become a politician – and his union describes him simply as a 'farming stalwart'.

I want to be like John. If a punk-loving Brummie skinhead in Dr Martens can make his forever home in a remote Welsh farming community, then surely I can figure out where I belong.

M Y PARENTS, AVRIL AND TONY, were married in 1979 at the eleventh-century Norman church in Llanyblodwel near the banks of the River Tanat, which flows from the Berwyn Mountains, north of Lake Vyrnwy, over the English border and into Shropshire. Each of them had relatives and ancestors buried in the ancient graveyard (so they're true locals) and most of their friends and family were already acquainted, or at least recognised each other from around the local area; most likely the livestock market, pub, or church. They were two young people with ties to their land and community as deep and binding as the roots of the old oak on 'The Hill' next to our small farm, just on the English side of the border.

The Hill, our hill, is a 25-acre field which Grandad Bill used to farm. It was sold at auction in 1991 after he retired, along with the rest of Craigllwyn Farm. This is where our family story diverges from British farming tradition – of farms being passed down from generation to generation or more likely, even today, from son to son. It's a sad story which illustrates perfectly the idea of a 'heaf'; that sense of belonging to land.

Despite farming at Grandad's side from boyhood and knowing every inch of the land where he was born and bred, Dad was never under any illusions that he, nor any of his siblings, would ever inherit it. For reasons I have never quite understood, other than there being some uncomfortable family politics around the early twentieth century, my great-great-grandfather, William Jones (son of that revered farmer of the same name) bought Craigllwyn for his 'kind but crippled' son John Wddyn. He decreed that the farm should provide a home for John's young wife Elsie, aka Nanny, for as long as she lived (she was expected to outlive her sickly husband even then). But upon Nanny's death, the farm must be sold, and the proceeds split equally between their children – Grandad Bill and his three siblings.

Nanny went on to rule over the farm for more than 70 years, dominating first her husband and later her son. Her cast-iron will made up for any physical strength she lacked as a woman. Sharp, business-minded and desperate to farm in her own right but feeling constrained and resentful of the male-dominated world within which she was forced to operate, Nanny railed against the prison of her gender. She turned not to feminism and the fight for equality but appeared to choose, instead, a path of bitterness and jealousy. I've heard Nanny described as cold, cruel and a relentless bully, particularly against younger women, including her own daughter-in-law (my grandmother) and her granddaughters (my aunts). But I also sense a shrewd and intelligent woman, with an unnurtured talent for the job she loved and vast amounts of ignored potential. What a terrible frustration for any woman to bear.

Nanny died at the age of 92, and Grandad Bill's life's work died with her. He had farmed quietly in the shadow of his autocratic farm partner, his mother, all his life and now she was dead, it was time to sell up and leave. Buying out his siblings was a financial impossibility. The family farm was lost.

The livestock and equipment were sold first on a cold autumn day in the late 1980s. The auctioneers arrived early on the day of the farm sale and went around the livestock, assigning lot numbers to each of Grandad's 25 Welsh Black/Hereford-cross suckler cows, his prized Charolais bull, about 100 Welsh ewes and even his chickens. The auctioneers brought their own portable cattle ring and set it up ready for the sale in the stackyard by the barn. Every bit of machinery and farm equipment, from scrap iron to tractors and trailers, was placed in rows in the field nearest the farmhouse. Around 200 people turned up for the auction; mostly neighbours and local farmers who customarily made sure they bought at least one thing in support of a well-liked, long-established local family. My Dad, who was in his late thirties at the time, bought a cow and calf and his father's Massey Ferguson tractor.

'It was a very busy day,' he remembers, 'There wasn't time for emotion until it was all over. We had supper in the house and talked about it a bit then, but he (Grandad Bill) was very quiet. He kept it all to himself, which was probably a bad thing.'

My grandparents' house was sold in 1991 to a young family from Manchester. Steve and Sarah and their two children Rachel and Tim became our neighbours and good friends. They were the first city dwellers, as far as I know, to move into our little community 'on the top'. Sarah Gibson, now divorced and a successful nature writer and conservationist, grew up in rural Sussex and always considered herself a country person, but life and work had taken her to Oxford, London, and Manchester. She remembers the day she viewed Craigllwyn farmhouse for the first time and met my Nan Beryl:

'She was the perfect person for selling a house. She'd made fresh scones, and the fire was going, and it was all so welcoming. Rachel came in and Beryl said, "Oh she's bright as a ribbon." I always remember her saying that. It was such a sweet thing to say.'

'I felt uncomfortable when we moved into Craigllwyn,' Sarah adds, sharing something I've never heard her say before, 'like we'd displaced Beryl and Bill and that we didn't really belong there. It must have been so hard for them to leave.'

The irony of their passing lives has only just struck me – as Sarah returned to country life after years in the city, my grandparents were about to move to a town for the first time. It was an urban/rural life swap.

The inevitable sale of Craigllwyn Farm was a painful, drawn-out affair that had hung over Grandad Bill all his life and finally culminated with the auction of his land. Somehow, rather unwisely, he had never reconciled himself to the unavoidable fact that he would, eventually, lose his beloved Craigllwyn. There is no doubt in my mind that the loss of his home, his work, and his identity, coupled with the subsequent move to a small, terraced house in Oswestry broke his heart. His share of the money from the sale of the farm could never compensate for what he'd lost, because roots meant more to him than money. Grandad Bill died in 1995 , aged 72.

But the umbilical cord between our family and Craigllwyn Farm was not completely severed. My Dad was able to cling on to The Hill – 25 acres was all he could afford to buy on the day of Grandad's land sale. He'd hoped to buy the full 65 acres and the farm buildings by clubbing together with my uncle, his younger brother Stephen, but they were outbid.

I've often wondered why Dad went to such lengths to buy The Hill and stay near Craigllwyn. He could have lived wherever he pleased – there was nothing keeping him there when Grandad Bill sold up and retired; it was the end of an era for our family. Perhaps it would have been easier for a man of 40 with three young daughters to move on and start afresh somewhere else. Apparently he did look, half-heartedly, at some council farms and even considered some farm tenancies but could muster no enthusiasm for anywhere but Craigllwyn.

My parents ploughed every penny they had ever saved, and a lot of money borrowed off the bank, into buying The Hill and a modest bungalow on adjoining land, which they purchased from an elderly widow in 1988. I often see the surprise register on friends' faces when they visit our small farm for the first time, perhaps even a tinge of disappointment when they discover it is not a stone cottage with roses around the door, or a fancy red-brick pile surrounded by timbered barns and historic National Trust-esque buildings – but a 1950s bungalow with PVC windows. It is not your typical farmhouse.

Dad says it was the only place he could even hope to afford to buy in the late 1980s/early 1990s, when interest rates were through the roof and the property market had gone stark raving mad with gazumping. While I'm sure that's true, I struggle to believe that's the whole story; especially knowing my Dad. I've watched him on a still midsummer's evening, standing on the highest point of The Hill just behind our bungalow, overlooking the cattle grazing in the hollow, bellowing softly under a blue sky washed with pinks. He will stand there, hands in his pockets, with a look of such peace and contentment, surveying his piece of land. His home. His heaf. 'They'll take me from here in a box,' he has said on more than one occasion.

I've heard many farmers talk like this. They possess an enviable serenity with mortality, safe in the knowledge their land and life's work will still be here long after they're gone. Sinead Fenton, the first-generation farmer living in a caravan on a four-acre field in Sussex, is only 30, yet, perfectly casually, she told me: 'This is probably where it ends for me – and not in a dark way. I feel quite content and happy with that. I don't feel scared about the idea of dying here. We want to leave this in a better state than what we found it in and hopefully we're on the way towards doing that.'

I'm sure my great-great-great-grandfather William Jones felt exactly the same when he was busy flattening his fields and straightening his hedges. Good farmers are like worker bees, dedicating their lives to the greater good of the hive.

Dad turned a single field farmed by his father and grandfather into a farm of his own. In 1993, he built a large modern shed with steel structures and Yorkshire boarding; one side for wintering cattle, one side for lambing ewes and a tower of hay and straw bales in the middle. He divided The Hill into two separate fields and fenced it with the help of his youngest brother, Simon. Dad plunged himself into serious amounts of debt, which only deepened throughout the crisis years for farming that followed – from the BSE or 'Mad Cow Disease' epidemic in the early 1990s to Foot and Mouth Disease in 2001. There were times when Mum begged him to sell up, to get a job as a lorry driver, but Dad's grim determination to cling on to his farm and the place where he belongs are, in my mind, the physical manifestation of the deep connection to land that is bigger than ourselves.

One side of The Hill is steep-sided and covered in gorse, just about walkable thanks to the narrow paths cut by cattle and sheep over the years. You can walk along the bottom, following the hedgerow, but it quickly turns into a mud-bath in wet weather and churns up under the cattle's hooves. When it's dry, their footprints bake into rock hard ruts and bumps – like walking over the top of a meringue. To avoid twisting your ankle, it's better to climb up the side of The Hill and find a sheep path through the gorse. Over the years I've learnt to contort my body like a limbo dancer to achieve safe passage through the needling bushes. Once you hit the stream, the jungle of gorse peters out and The Hill gets grassier and rockier with grey stones scattered here and there.

The other, more intensively farmed, side of The Hill is permanent pasture and has never been used for anything other than grazing cattle and sheep. It was ploughed and reseeded once by Grandad Bill in 1947. It overlooks Llansilin and out towards the Berwyn Mountains beyond, which lead all the way to Snowdonia. There's a great view of Gyrn Moelfre, a 523-metre summit which we know simply as 'The Gyrn' (pronounced 'gurn') It featured in the 1995 film *The Englishman Who Went Up a Hill but Came Down a Mountain*, starring Hugh Grant. The fact I still pull this out as a favourite piece of local trivia gives you some idea of how exciting that was.

Look out towards the Gyrn from The Hill, let your eyes fall to the foot of its south-western slope and there lies the little village of Llansilin, population 698 as of the 2011 census. It twinkles like Bethlehem on a December night. I feel like a shepherd looking down at the tiny specks of light in the darkness below, to a village community that once felt like an extension of home.

My grandfathers, father, my sisters and I all went to primary school in Llansilin, as did dozens of our cousins, aunties and uncles. Up and down the twisty-turny road to Llansilin we'd drive, throughout the 1990s and 2000s, seven days a week, for school, church and Sunday school, Brownies and Girl Guides, concerts, shows, Eisteddfods, plays, parties, pantomimes, bingo nights, whist drives and WI.

Secondary school was further away, 50 minutes on a bus which left at 8 o'clock sharp from outside the village hall. We were always late. Poor Mum, red-faced and stressed as she trailed the bus down the road, willing Merv the driver to spot her in the rear-view mirror and pull over. He always did. I'd jump out – 'bye Mum!' – grab my Umbro rucksack and squeeze between the bus and the hedge, picking my way along the muddy verge, school shoes leaving footprints in the soggy tyre tracks of tractors. As the bus pulled away, I'd watch Mum do a three-point-turn and head back up the windy road to home; to get my younger sisters ready for primary school. She'd be back in Llansilin within the hour. 'I spend my life on that ruddy road,' she'd cry, desperate for help, though Dad rarely had time to take us.

Our school days played out against a backdrop of busy farm life. As I trundled back and forth on the bus each day, worrying about homework, friendship quarrels, or boys I fancied, Dad was desperately trying to make money buying and selling sheep in the livestock markets during one of the most challenging times for British agriculture. In 1997, the year I sat my GCSEs, Dad had one of his worst years. Already crippled with debt from the purchase of the bungalow, The Hill and the building of the shed, the economic downturn in farming threatened to crush him.

'The worst years were 1996, 1997 and 1998,' he remembers. 'The markets, trading, exports – everything just collapsed under our feet.

Everything went bad. When you get one bad year you can bounce back. But when it's three continuous years ...'

I remember Mum trying to save money on the housekeeping. As a full-time housewife, she relied on the cash Dad gave her every week to feed and clothe us. Once, she bought cheap breakfast cereal that tasted like cardboard. Dad was appalled. 'Never let your stomach know your pocket is empty,' I remember him saying.

In their darkest moments, Mum would beg Dad to quit and sell up. She's often intimated that, as a butcher's daughter, she could never quite get her head around Dad's single-minded determination to persevere with farming:

'In 2001, during the Foot and Mouth, it was the first time I saw him crying. He was sat on the settee and said, "It's all slipping away, and I can't stop it." We hadn't got any money, we owed feed bills and things, so I got the number off the Teletext and called the Farmers in Crisis helpline.'

They put Mum in touch with the Royal Agricultural Benevolent Institution (RABI) who sent out a cheque for £2,000 so we could pay our bills. I say 'we'; I had no idea any of this happened. The first I'm hearing of it is today, sat at the kitchen table with Mum, 20 years later.

I never once felt poor. The signs were there I suppose – most of our clothes were hand-me-downs from cousins in town and we missed a few school trips and weekend activities with friends. 'Oh come on, just get your dad to sell a sheep,' they'd plead. That sounded simple enough, so I'd go home and ask him, to which he'd shake his head and sigh: 'It doesn't work like that Anna.'

I was my parents' first child, born in 1981, and the first in our direct family line, stretching right back to Richard Roberts, that pauper agricultural labourer born in 1777, to go to university and move to a city. Before that, only war had come between my family and a quiet rural life. Grandad Wilfred served in North Africa with the RAF from 1939 to 1945. He returned home without a scratch, picked up his butcher's apron, married my Nan in 1947 and vowed to never again leave the UK, or the Welsh Borders if he could help it.

As much as Grandad loved being a homebird, I yearned to fly away. I knew from a young age that I would leave; right from the time I clip-clopped around the farmyard in Mum's heels, dreaming of concrete and skyscrapers.

I wasn't the only one. There were nine pupils in my year group at Llansilin primary school. All but one left, most of us to university in large towns and cities.

This a familiar picture in rural areas right across the UK – young people leave, they get jobs in the city, they settle elsewhere and rarely return. My disconnection, at the age of 18, from the land to which my forefathers and mothers belonged, in search of a busier, brighter destiny in the city, mirrors the story of the wider UK population, over hundreds of years. The story of our Great Disconnection.

And it's the same story all over the world. Follow your own family tree or dig deep enough into your DNA, and you will find a farmer. Most European men are descended from the early farmers who spread westwards from the Middle East some 10,000 years ago, displacing hunter-gathering with their crops and animals. In fascinating contrast, most female lines descend from hunter-gatherers, suggesting Neolithic women chose the newly arrived agriculturalists over indigenous hunter-gatherer men. There was a time when being a farmer was very sexy indeed.

And that's the way life went on for millennia. The whole world was rural. Most of us were farmers. It wasn't until the 19th century when two revolutions, industrial and agricultural, sowed the seeds for disconnection.

Subsistence farming, which had kept people on the land for centuries, gave way to new, professionalised methods of production. Horses were replaced with machines and, gradually, so were the farm workers.

With more food being produced, and fewer hands needed to do it, people were free – or forced – to pursue new ways of making a living. The general population shifted from relying on the land to produce their food, to relying on the farmer to do the job for them. A responsibility once shared by the many was handed to the few.

Most of us left the countryside; migrating en masse to the factories and mills of Britain's expanding towns and cities. As the nation shifted from an agricultural economy to a manufacturing one, the people changed too. Sons and daughters of farmers became the parents of city dwellers. This was the beginning of the Great Disconnection.

Of the 67 million people living in the UK today, just 17% live in a rural area and less than 1% work in agriculture. The total agricultural workforce, which includes forestry and fishing, is 472,000. More people live in Edinburgh than work in farming throughout the whole of the UK.

What we are talking about here is a minority group.

Farmers and country people talk to me about the disconnect – the urban/rural divide – all the time. It comes up in conversation constantly. They are much more sensitive to it than urban folk and spend time ruminating over it, and consciously feeling it. The person who best summed it up for me is a young American farmer called Brandon Pickard.

He runs a large-scale arable operation in rural Iowa with his father, uncle and younger brother, Ryan. It's a monoculture of corn and soybean, heavily dependent on the chemicals and fertiliser purchased from their local agronomist and farm service based in nearby Melbourne, a city with less than a thousand people.

When I first meet Brandon in 2018, he's struggling to pay the bills as President Trump wages his trade war with China and grain prices tank. He's losing tens of thousands of dollars on corn and soybean.

'It's not a good feeling at all,' he says, 'when you're sitting here in a $400,000 machine and you gotta make the repayments on it. You hope you make enough to cover all the repayments on everything you've got and hope to put a dollar in your pocket.'

The annual costs of running their 5,000-acre farm are eyewatering. Everything gets split four ways and Brandon, who always does me the honour of total transparency, breaks down his rough share of the costs:

Seeds (genetically modified field corn and soybeans): $500,000

Pesticides, herbicides, and fertiliser: $400,000

Rent: $200,000

Machinery repairs: $120,000

In a good year the farm should generate a total income of around $4 million, of which Brandon receives a quarter. After bills and debt repayments, his net salary for 2019/20 was $20,000.

'I'm probably making as much as someone on minimum wage,' he says cheerfully, with no hint of resentment at all.

Bizarrely, he makes a much healthier profit on his hobby sideliner – the six cattle he rears and fattens at home and then sells direct to neighbours and friends in beef boxes:

'My wife and I crunched the numbers and worked out that if I could get 10 to 15 head of cattle, we would make the repayments on the house.'

Brandon, in his jeans and baseball cap, looks like any other Midwestern row crop grower. Critics of 'Big Ag' or 'industrial farming' might write him off as just another chemically addicted, soil trashing, climate killing farmer, without much thought for his story or character. Brandon belongs to a rather faceless group of people, often lumped together and talked about in the news or at conferences as 'conventional farmers' running large scale, industrial systems. He and Ryan recently joined the ranks of 'factory farmers' after investing in two hog barns, each housing up to 5,000 pigs. There's very little mainstream discussion about the individuals who belong to this group.

Brandon is a deep thinker and an intelligent conversationalist, with kind eyes and a gentle demeanour. He's a family man and a farmer to his bones.

There's no doubt Brandon is fiercely ambitious, measures success in the scale of his farming operation and openly admits he wants more and more, and to farm bigger and bigger. He doesn't believe modern agriculture has had a negative impact on the environment, but that's not to say he doesn't care either. When combining his crops, planted intensively right up to the creeks and roadsides, he'll voluntarily leave patches of long grass and unharvested crop for the wildlife to run into and shelter.

From the cab of his red Case IH combine, I spot a couple of pheasants and a hare darting in front of us, bolting for cover. 'I like seeing them,' he smiles.

I first get to know Brandon during the 2018 soybean harvest when he agrees to chat to me for a documentary series I'm making for the BBC World Service. I sit next to him, in his air-conditioned combine, chatting as the machine rolls through a vast 70-acre field of beans. The floor of the cab is carpeted, 'to make it more homely because I spend so many hours in here'. If there's rain forecast, which threatens to damage the crop, he's been known to spend 18 hours straight combining, hardly seeing his wife and two children. Sometimes he doesn't even have time to take a leak.

His Mum drives over to meet us with a packed lunch, flagging us down from the roadside, and hands us a bag of lovingly prepared sandwiches, cake, fruit, and water. Brandon pops the bag of goodies into the little fridge beside his seat. I ask him: 'How do you think urban people view you and your way of life?'

His answer will stay with me forever: 'We're out here just playing in the dirt. We're just the Flyover States. No one even really knows or cares that we're out here. We're nothing and yet, we're everything.'

Those words were epiphanous for me. Matter-of-fact eloquence, spoken at the wheel of a lonely combine in a vast landscape which, to me, looks as empty and desolate as the surface of the moon. In that moment, Brandon sums up what farmers and many rural people mean whenever they speak of the 'disconnect'.

I happened to mention it to a well-known New York food justice campaigner called Karen Washington. Her work is centred around getting fresh, affordable, organic produce into deprived areas of the city. I want to get an urban perspective on what Brandon told me.

'Farmers . . .' she shakes her head, 'these are people who are doing so much for us as Americans but yet get little respect. I would love to meet that farmer and shake his hand and tip my hat to the fact that what he's trying to do is just feed his family and feed the country.'

I wasn't expecting such sympathetic solidarity from an agroecological, organic, urban, market gardener towards the Midwestern factory farmer – but why not? Why shouldn't they stand together? Who says these two worlds should be divided? And what has division achieved anyway?

Whether or not urban people ever stop to think about individuals like Brandon is by the by. The fact many rural people feel invisible is proof enough that the disconnect exists. There's a deep sense that they are unnoticed, ignored and forgotten. And yet the survival of our species is utterly dependant on them. It is an inarguable fact that we, all of us, simply could not survive without the rural bedrock upon which the rest of modern, urban life is built. Only full-time hunter-gatherers or foragers can feed themselves without the service of farmers and growers (and perhaps allotment owners and gardeners, who are, in effect, farmers.) You cannot separate your existence, as a human, from agriculture.

But this isn't just about farmers – it's the whole food system. The complex, predominantly rural, networks which revolve around the production, processing, storage, manufacturing, retail, and delivery of our food. Vets, mechanics, agronomists, contractors, hauliers, engineers, technicians, consultants, builders, livestock markets, auctioneers, abattoirs, processors, factories, land agents, grain merchants, animal feed suppliers, farm workers, herdsmen and women, tractor drivers, foot trimmers, shepherds, shearers, lorry drivers, forklift truck drivers, rural specialist solicitors, surveyors, accountants – and all the rural services that support their life and work in the countryside – shops, banks, Post Offices, pubs, cafés, schools, libraries, doctors surgeries, dentists, hairdressers. I'm not saying for a second all rural life revolves around farming – many people living in rural areas would argue they have no personal or professional connection to the agricultural community whatsoever – but if you were to take that bedrock industry away, a vast, interconnected rural system, a chain of livelihoods you never even knew existed would come crashing down – and the shockwaves would hurt us all.

It is a parallel world that supports and enables our lives in the city.

In my view the disconnect does not come from the fact we take all this for granted. I'd challenge anyone who says they feel perpetually thankful and grateful for food; or somehow expects consumers to throw themselves upon the altar of agriculture every time they eat a pie.

No, the disconnect Brandon described is much more subtle than that. It's about mislaid preconceptions about who rural people are and how they live. It's about assumptions made about their communities, widespread ignorance of their needs and the metropolitan instinct that ruralism is less relevant to the things we place great importance on: government and politics, business, economics, services, culture, convenience, choice, and simply 'being busy'. Few would disagree that 'the countryside' is a lovely place to visit for a holiday, escapism, exercise, soul searching and spiritual renewal. But for living actual, real life? Culturally it's a backwater, politically it's irrelevant, economically it's lacking in opportunity, the services available do not meet my requirements and the people there do not share my world view.

While we may not consciously think these things, or all at the same time, they are real. I'm ashamed to say I've had some of those thoughts. I'm ashamed it only took one generation of separation to view the rural way of life as 'other'.

In my view, the disconnect can best be described as collective memory loss. We have simply forgotten where we all came from.

CHAPTER TWO

WORK

A FARMER NEVER SWITCHES OFF. Growing up, Dad was always working. Not hidden away in an office, or constantly on the phone or in meetings – like workaholic executive dads in movies – no, he was always around; it was just his jobs were never, ever finished.

Dad's work was also our holiday time, spare time, and family time. In fact, we never even called it work – it was always referred to, rather reverently, as our 'way of life'. An all-encompassing existence. There was no beginning or end to the working day – it was just a day. As a result, I grew up with the very distinct impression that only town people 'worked' while we had a 'way of life'.

My happiest childhood memories are of climbing up into the cab of our stock lorry with Mum, Dad and my two younger sisters, Kate and Nicola, with an enormous picnic. We'd drive down-country to the lowland grazing Dad rented on the banks of the River Severn near Shrewsbury and spend our entire Sunday sorting fat lambs for market on Monday or weaning lambs or tailing ewes (trimming the pooey bits off their bums so they don't get maggots in the hot weather).

Sometimes, maybe once a year, we would take a day trip to the beach in North Wales. We had one foreign holiday to the Algarve in 1993. But mostly, family fun, relaxation and recreation were all tied up with farm work.

Of course, this is less about living rurally and more about being farmers. It is a job unlike any other, as Leicestershire dairy farmer Ruth Grice discovers more each day. She is transitioning from an office-based marketing and fundraising career in Oxford and Nottingham to becoming a full-time dairy farmer near Melton Mowbray.

Ruth's decision to quit city life and work alongside her father on the farm, which has been in their family for four generations, first struck me as wonderfully spontaneous, utterly inspirational and a perfect back-to-the-land story for this book. She's the Diane Keaton character from my childhood daydreams! Ditching the stilettos and briefcase, turning her back on the rat race and marching out of her office and straight into the milking parlour, where her stoic father, who secretly always hoped his daughter would return, silently weeps with happiness.

Of course, it's nothing like that because Ruth is a real-life person (who has never worn stilettos or carried a briefcase in her entire 11-year career at the Wildlife Trusts). In reality she is creeping back to the farm and cautiously edging her way into the family business which, incidentally, had been getting on perfectly well without her. The transition from office to milking parlour has been carefully orchestrated to reassure Ruth's extremely dubious parents William and Jane who, far from welcoming their daughter home from the city with open arms, subjected her to a harsh interrogation:

'They really challenged me as to whether it was the right decision,' says Ruth. 'There is so much I don't know and so much I can't do, I guess they were challenging what I could bring to the business. On the financial side of things, they were concerned about me being out of pocket because my salary would be different.'

Ruth had to prove herself. She started off working just one day a week on the farm in 2017 while continuing in her marketing role for Nottinghamshire Wildlife Trust part-time, but her future direction of travel was obvious. In November 2020 she went up to two days a week and openly admits she plans to quit her marketing career altogether. Her cautious approach is paying off:

'It's taken me a good three years to find my niche in the business,' she says. 'I love a good spreadsheet and I love my data. I had not appreciated until I came back to the farm how data-driven the industry has become. The bi-weekly vet checks that rely on data input and data analysis, making sure the cows are sticking to their lactation periods – Dad's got it all in his head; he doesn't need a spreadsheet to be able to know what's good and what's not. I don't have that knowledge, but I do have a very good spreadsheet that can help me. I am helping with that side of the business, and all the paperwork and accounts and financing. I can do it. I know I can add value to the business on that front.'

Far from leaving office life behind, Ruth has brought the skills she acquired in the city back to the farm. She's still sat at a computer updating spreadsheets – but something else has changed. Ever so subtly, Ruth is undergoing the transformation from a woman who works, to a woman with a 'way of life':

'I will happily do my cow data on a Sunday evening in front of the telly and it's fine. I don't think of it as being work. But I would never do that with my Wildlife Trust work – I'm much more 9 to 5 with that.'

Checking herself, she quickly adds: 'Don't get me wrong, I still work hard – it's just . . . different.'

In farming, unlike most jobs, you don't clock off, go home, and forget about it until tomorrow. The farm is your home – the last thing you see before you go to sleep is waiting for you when you wake up. I believe it's the main reason farmers take criticism of their industry so personally and get extraordinarily defensive in the face of direct challenge – livestock farmers in particular. They spend every day of their lives caring for their animals, they put the needs of those animals before their own and brave all weathers to make sure they're fed, watered and safe. Many critics of the industry will roll their eyes at that – smelling another sob story pedalled by self-pitying farmers – but, really, we have no right to question it unless we have done the job ourselves. For years. And for little or no money. I'm a farmer's daughter and, oh, how I wish I could do what Ruth has done. I admire her so much. But I don't have what it takes – you probably don't either. Most of us were not born with the X factor needed to farm; to live the 'way of life'. We should respect those that do.

But there's a flip side to this deeply ingrained existence – this institutionalisation. Throughout my childhood and career as a rural affairs journalist, I've listened to farmers revelling in their martyr-to-work mentality, boasting about how many weeks and months they've worked without a day off and generally wearing exhaustion like a medal. What's more, without even realising it, I've inherited this mindset and taken it to the city. It's created some domestic discord in my life and even caused a few issues in relationships, certainly as I've got older.

My partner Alex and I have been locked in the same entrenched debate for most of our six-year relationship – do we stay in the city or move to the countryside? He loves Bristol. I dream of moving home to Shropshire. There is nothing new about this classic tug-of-war – most couples will encounter the 'where do we live?' conundrum at some point – but our enduring stalemate feels particularly gruelling. It's like we are fighting for our very identities. Alex is an extroverted socialite who feels recharged and energised by Bristol's thriving music and arts scene, its diverse pubs, bars and restaurants and the numerous outdoor events and festivals which pack out the city's parks and green spaces every summer. Don't get me wrong, I love all this stuff too – I'm joyfully by his side – but far from filling my cup, it drains my energy levels. To recharge and reenergise I need wide open spaces, fields, hills, mountains, peace and quiet. Alex will joyfully join me for hikes, camping trips and holidays on the farm – he also loves these things – but it doesn't fill his cup. Too much quiet time in the countryside leaves Alex listless, lethargic, and lonely. He needs people and bustle. If we didn't love each other it's unquestionable our own personal urban/rural divide would have broken us up long ago.

I have always dated 'townies' and city boys, and by far the most difficult part of an urban/rural cross-cultural relationship is our conflicting definitions of downtime. I struggle to switch off from work; I find 'sitting around' a waste of time; I feel guilty about being indoors when I could be outdoors. This is without doubt symptomatic of growing up on a busy working farm. Gethin Bickerton, the farmer's son from my neck of the woods who's now a Cardiff-based actor totally gets it:

'I've been complimented in Cardiff and other cities on my work ethic, and I can only put that down to a life in agriculture and growing up in the country. I am not one to sit down and do nothing. I will feel incredibly guilty because, growing up, if you were sitting down watching telly, well, you could be outside feeding the sheep, couldn't you?'

I am hardwired to prioritise work over life. I will put off jobs around the house for months, even years. For a large chunk of my childhood, a screwdriver knocked into the wall held up the curtain rail in the living room and my dead great-grandmother's walking stick held up the grill on the oven. Our old galley-style kitchen quite literally rotted away. Yet the animals always had food, water, and fresh bedding. They came first.

Sometimes it would get Mum down and she'd kick the walking stick in frustration and threaten to go and live in a caravan at the seaside unless Dad bought her a new kitchen. But her fury would pass, and the walking stick was dutifully returned to its post.

Work dominated everything. I can't remember a single occasion when Dad said 'yes' straight away to a social invitation. Not one. His reply would always be: 'No,' or 'I doubt it.'

Sometimes, on rare occasions, we'd get a tantalising: 'We'll see.' Because there would always be a cow calving; or a ewe lambing; or hay to cut; or any number of reasons why we couldn't go. It was far better and simpler to just not make any promises at all and avoid letting people down. That made life a lot easier. It's a philosophy I have kept going through my own life.

I struggle to commit to social engagements until the very last minute (seriously, I can be a right miserable git: 'I haven't got time for fun! Don't you know how busy I am?'). I consider a weekend at the farm a holiday – whether that's helping Dad shovel beet out of the back of a trailer or squelching through muddy fields erecting electric fences. To me this is fun.

Alex does not find it fun. Not all the time anyway.

Everything I have listed is a source of endless frustration to him. He's a Leicestershire lad who grew up in a very nice semi-detached house on a main road in Loughborough. 'You never stop!' he says, on pretty much a weekly basis.

One day, as we sit opposite each other at the kitchen table quietly working away, I decide to ask him about it.

'Where do you think it comes from?'

'Well,' he ponders his reply, 'Your mum never stops either, even when she's not that busy. And coupled with your dad's work ethic as a farmer, I guess you can't escape it. It's built in you.'

I ask about his childhood, growing up in the town – did his family 'stop'?

'Yeah, we'd plan things for our weekends – we'd go camping or on bike rides. Mum always said Sunday was our time. We had a roast dinner, sat down at the table, and had a good old get-together. There was a real sense of occasion. Sundays were sacred. But it was different for me – my parents were teachers who did 9-to-5 jobs. When it got to Friday, they were going to have a weekend because they'd had a hard week.'

I mull this over. My upbringing was different in almost every way.

I can't remember my Mum sitting down in the middle of the day to have a cup of tea, read a book and relax. Ever. She did everything single-handedly – cooking, cleaning, shopping, the school run and the lion's share of parenting three children (and hundreds of pet lambs). In 1997 our farm business was struggling so desperately she went out to work, first at a local corner shop and then at Argos in Oswestry where she stayed for 16 years. She worked weekend shifts as it was the only time Dad was at home to keep an eye on Kate and Nicola, as weekdays were market days. I wasn't around much either as I had various Saturday and Sunday jobs from the age of 13. All in all, weekends were nothing special for our family – we just kept on working. I don't lament this fact. I enjoyed my weekend jobs, Mum loved working at Argos – it gave her a break from the farm and us kids. And my sisters? Well, they got to watch inappropriate television while Dad worked outside. 'Don't let them watch my Jilly Cooper's *Riders* video!' Mum would holler down the phone during her lunch break.

We were a happy, busy family who whistled while we worked, but Sundays lost their sacredness due to necessity.

It wasn't just our family. I've seen this 'way of life' all over the UK and around the world. Farming families live season to season, muddling through each year as it comes and goes. Some years are good, some are bad. The rhythm of our life was set by the metronomic constancy of lambing, calving, muck-spreading, shearing, harvest, tupping, sorting sheep, selling sheep, and buying more sheep.

While I've always worn my country girl identity with pride – praise be the rural way of life – this deeply ingrained, institutionalised mindset is far from perfect. As well as being mildly annoying to our friends and loved ones, it can also be selfish and lacking in empathy for other, different ways of life. And I believe it has contributed to the urban/rural divide.

Many farmers, including my own dear Dad, behave like they have a monopoly on being busy. They seem to think urban folk, with their 9-to-5 jobs and weekends off, have an easy life mowing their neat lawns, and washing their cars and going shopping. I encounter this attitude frequently in farming circles. I've lost count of the times I've been reminded, usually during research calls for the BBC, that 'the countryside is a working landscape you know!' The insinuation being that day trippers and tourists from the towns and cities, with all their free time and hobbies, are just an annoyance, 'getting in the way of our work'.

I am coming to realise that living and thinking like a workhorse and martyring yourself to a 'way of life' is flawed and flies in the face of all the best advice around achieving work/life balance and protecting mental health. And some farmers are reaching the same conclusion.

Robert Thornhill milks 280 Jersey/Friesian crossbred cows on a pasture-based system in the Peak District National Park. The herd grazes a pretty patchwork of fields a few miles north of Bakewell. The boundaries of each paddock are marked out with a mixture of electric fences and dry-stone walls and old ash trees grow in single file along the edge of fields, where they seeded themselves many decades ago. Rob's grandfather was a poultry farmer and the cows arrived in 1954 when his father took the business in a new direction.

My first visit to the farm was during the hot summer of 2018, at the height of that year's drought. Rob had lost weight since I'd last seen him at a farming conference the previous autumn. He looked stressed; his face pinched with worry. I only had to look at the scorched ground to understand why.

Rob's whole farming ethos, which largely emulates the New Zealand dairy system, relies on grass. It is meticulously measured, intensively grazed, and carefully managed in a strict rotation. When it doesn't grow, there's nothing for the cows to eat and no hay or silage to see them through the cold winter months ahead. For the first time in his career, Rob was forced to buy in a cereal-based feed called alkalage, a kind of silage made from barley. It more than doubled his cost of production and severely knocked his confidence. For Rob, pasture-based dairying is more than just a method of production – it's a philosophy. In 2013 he travelled the world studying forages and grazing techniques, his own personal pasture pilgrimage. To end up resorting to cereals felt like a betrayal to himself.

To top it all off, in that cruel summer of 2018, the farm was short-staffed. Rob was also building new winter housing for the cattle and was drowning under mounds of office work, which increasingly characterises modern farm management.

So, Rob knows hard work. He knows stress. I can see in him the potential for workaholism – and it's a compulsion he has worked hard to resist. To maintain a healthy relationship with his farm and his cows, Rob takes holidays. He has hobbies. I don't know anyone with a longer list of hobbies than Rob. He goes skiing, sailing, motorcycling, target shooting, scuba diving, running, and cycling. He laughs when I rattle them off:

'And I'd do more if I had the time, Anna! I'm very open to infection from people's enthusiasms. I shared a table at a wedding with someone who was into skydiving, and I got the bug. When you undertake other activities and hobbies you mix with non-farming people which is so refreshing and very invigorating for the mind and gives you total switch-off. When I'm squeezing the trigger on a long-distance rifle range, I'm not thinking about the level of grass my cows are eating.'

He vociferously rejects the 'farmer martyr' attitude to work and believes many are damaging themselves and the industry by never leaving their farms. I wonder out loud if this, far from being an act of self-sacrifice, is an act of selfishness? He nods swiftly:

'I think it is selfish because you're putting your work above everything else. There's often a feeling of entitlement, like: "It should be recognised that I haven't missed a milking in 40 years – I am doing it for you, and I am great!" I think that's a terrible attitude.'

I must look doubtful, because Rob continues: 'Many farmers do think that, certainly dairy farmers, and they're grumpy as hell. So I say, "Why do you do it then?" "Well, I love it," they say. So why are you moaning? If you like it – love it and embrace it. If you're grumpy – get out and stop moaning. Nobody is born a farmer. Nobody is born anything. You do everything by choice.'

Rob credits his father for teaching him to play as well as work, and said they always had family holidays growing up in the 1960s. This strikes me as unusually fortunate for a generation that farmed through the 1950s – particularly when I compare it to Grandad Bill's working life at Craigllwyn Farm.

'I'm very much of the opinion that you've only got one family to look after,' says Rob. 'How big a business do I need to achieve what I want, which is to look after my family comfortably and to be able to have nice holidays?'

I wholeheartedly agree with his logic – and applaud it – as I've long balked at the increasingly expansionist mindset in agriculture. The obsession with getting bigger and bigger, working harder and harder, with higher and higher costs – financially and environmentally – for what is very often minuscule extra profit, and a lot more stress.

Sometimes, though, overwork isn't driven by ambition – it's about survival. A sudden memory pops up from my own childhood – of peeping round the door to my parents' bedroom as they tip out a collection of 20 pence pieces from an old whisky bottle. Together, they count out the coins on the bed.

They needed the money to pay for a school trip for one of us. At the time I marvelled at the pile of cash – about £60 – believing we were rich beyond compare. Now, those child's eyes long gone, my heart aches at how bad things must have got, and swells with love knowing how hard they worked to make it OK.

When Dad worked himself into the ground throughout the 1990s and 2000s, it was out of necessity. He was on the brink of losing everything. No matter how much his mental health needed it, nice holidays were simply not an option. Try explaining that to the bank manager.

But regular time off – even just a day, or half a day, every week – maybe that could have been achievable? Rob thinks so:

'Farming never stops and you can fill every hour of your life if you choose to. I think a lot of farmers either are, or think they are, poor managers of people; they're difficult to work with so they don't get people in to help them. They tell themselves they can't afford it but, in my opinion, I think most of them can't afford not to because they work themselves into the ground. There's no way you can be as productive day in, day out doing those sorts of hours. You never get a chance to stand back and analyse your business.'

I smile when I think of the times we've pestered Dad to hire some extra help. Always a futile mission: 'By the time you've shown someone else how to do it you might as well do it yourself,' he says. So, it never happened – unless you count my cousin Mark, who helped him out on Saturdays throughout high school. They grew very close, and it probably did release Dad, in some small way, from the relentless pressure of never-ending work.

As a family, we were open and vocal about Dad's addiction to farming – we'd moan about it and tease him and, though it never changed anything, at least he got to talk about it. The worrying times were when he went quiet and withdrawn and spent hours working alone in the shed on dark winter nights. Those were the times Mum feared the most:

'One of his friends killed himself during the Foot and Mouth,' she remembers. 'He already had mental health issues and it just put the tin hat on it – he went up to his shed and hung himself. He was a big friend of Dad's and that upset him terrible. He was a nice man. Dad just walked out of the house and walked over to the shed. I was worried about what he might do.'

The relief I feel that Dad pulled through those hardest of times is almost overwhelming. Especially when I think of those who don't make it. Most of us know, either instinctively or having experienced it ourselves, that isolation and loneliness often come as a set, with depression thrown in for free sometimes. Yet despite the seemingly obvious risks, farmers have, traditionally, been hopeless at self-care.

The statistics speak for themselves. In 2020, most of the calls made to the Farming Community Network's support helpline, a charity with 400 volunteers across England and Wales, related to mental health. A Farm Safety Foundation survey found 88% of farmers under the age of 40 rank poor mental health as the biggest hidden problem facing their industry. And my eyes fill with tears when I see the official suicide figures for 2019/20. They're higher than I expected: 133 people working on British farms and in associated agricultural trades took their own lives in a single year, and most of them men. Scrolling through the statistics, it strikes me that being male – not necessarily a farmer – appears to be the greatest factor. Suicide rates are worryingly high among male construction workers, roofers, plasterers, painters and decorators, factory workers and machine operatives. Data from the Office for National Statistics (ONS) said men accounted for about three-quarters of suicide deaths registered in 2019.

This isn't a rural issue. I, like millions of others, have fought my own battles with mental health. I endured crippling anxiety in my late twenties while living and working in Birmingham. Back then, when a crowded pavement or a boy racer revving his engine was all it took to trigger a panic attack, I'm not convinced the mental perils of a peaceful rural life would have garnered much sympathy from me. I barely gave farmers a second thought. Why should I?

It wasn't until a conversation in the car park of a small-town American bar, on a chilly October evening in 2018, that I started to understand why farmers have a uniquely challenging relationship with work and mental health.

'We bought in, whole hog, to the idea of the rugged individualist,' says Wade Dooley, a sixth-generation farmer of 1,000 acres in central Iowa. 'We are the ones who do everything ourselves. We don't ask for help unless we absolutely must. For mental health, we never ask for help.'

Country and western music drifts out from Woody's Roadside Tavern where, according to the T-shirts for sale behind the bar: 'Drinkin' fast = thinkin' slow'. It's late on a Friday night in Albion (population 525) but we're both sober. I'm driving and still officially working, recording interviews for the BBC, and Wade is up early for another long day in the combine cutting corn and soybeans. The distant whir of his grain bins drying out the day's corn harvest hums through the quiet streets of Albion. He'll get a complaint or two about the noise from disgruntled residents, but complaining is something he's used to:

'We complain a lot as farmers in Iowa. We bitch about everything – the weather especially, that's our favourite. The markets, the economy, politics, soil conditions, the crop – whatever it is we're gonna bitch about it and we'll feel better after doing so.'

But if all the complaining doesn't make you feel any better, then that, according to Wade, is a bad sign. It means there are other deeper issues you've got to take care of. Something he realised himself when he started therapy, 'to help me deal with the depression I've not been dealing with.'

I shudder in the cold autumnal air. Wade immediately takes off his jacket and drapes it over my shoulders. I'm surprised – and, annoyingly, a little too giggly at this unexpected show of Midwestern chivalry. The thought strikes me that offering a lady his coat comes more naturally to this 'rugged individualist' than looking after his own mental health.

'Our culture has definitely been against mental health

reconstruction,' he shrugs, 'and so I have to do something because the future is very bleak for me if I don't. So now I'm going to therapy which . . . oh God . . . that sounds so horrible . . .'

He looks away from me, across the car park, into the dark night beyond.

'It means I have to admit I have a weakness that I cannot handle myself and that tears me apart inside, but it's kind of a release to say: Look I have a problem and I know what it is, and I need help.'

Therapy has given Wade some practical coping mechanisms that help him manage his depression:

'Being social on a fairly regular basis helps a lot,' he says over Zoom, two years after our chat in the pub car park, 'and making lists of tasks that I can tick off – even the small tasks that help me feel I'm accomplishing stuff – that helps the workaholic mentality.'

He disappears off the webcam and returns a few seconds later with a clipboard and several A4 sheets of paper crammed with handwritten notes.

'So, this is my list . . . and it's actually longer than that.'

If Wade feels himself sinking into a deeper depression – something he's learnt to recognise – it's time to roll out the big guns:

'I've got one of those stupid mindfulness books which you write in at the end of the day. When it gets really bad, I'll pull that thing out – I hate it, I've always hated journaling – but it does help. And the other thing is standing in front of the mirror, giving yourself a pep talk. It works but you have to do it for days in a row. It feels like a foolish activity but, again, it helps.'

Wade isn't the only farmer learning to ask for help. Rob Thornhill, the Derbyshire dairy farmer, signed up for mindfulness classes:

'In the first week it tells you to try and break your routine – analyse how you get dressed, notice which trouser leg you put on first, brush your teeth with the other hand. It's to try and reinvigorate a certain part of the brain that becomes habitual.'

'The benefits of training the body are widely accepted,' he adds, 'but less so are the benefits of training the mind.'

For all his hobbies and holidays, Rob, like many of us, struggles with stress management. He joined a mindfulness course called 'Focussed Farmers', specifically aimed at those working in agriculture.

It was set up in 2017 by Holly Beckett, daughter of the 'Beckett's Farm' family dynasty who have been farming on the edge of Birmingham for 80 years. They've made the most of their urban fringe location with a successful farm shop, restaurant, cookery school and conference centre. Derelict chicken houses, remnants of a past life in large scale egg production, have been converted into industrial units and offices for rent. The family continue to grow more than 1,000 acres of arable crops – a sizeable operation – but any mention of farming is buried on the website under a long list of awards for hospitality, food, and drink and 'The Best Breakfast in the Midlands'. These days, the jewel in their crown is people.

And it was studying people that led Holly to discover mindfulness in New York City.

Holly is a Nuffield Farming Scholar. As am I, and so is Rob actually. It's a travel study programme where you apply for a sponsored bursary to head off into the world researching a topic close to your heart that's related to agriculture. Afterwards you're indoctrinated for life into an international network of alumni.

My topic was the coverage of farming in the global mainstream media, which no doubt sowed the seeds of inspiration for this book. Rob went on his pilgrimage to pasture. Holly studied emotional intelligence and developing 'EQ' as opposed to 'IQ' – a quality often far more valuable to agribusinesses than academic achievements. And this is how she ended up on a two-day mindfulness course in New York with a group called the 'Search Inside Yourself Leadership Institute.'

'I never even thought about mental health to begin with,' she says. 'It was about leadership and self-development. But over these last five years I've realised that mental health is directly correlated with leadership. Mindfulness is about developing a higher degree of self-awareness and the best way to increase your mindfulness is through meditation or, as I call it, mental training, mental exercise, or brain training. I try to avoid the word "meditation" because farmers switch off straight away.'

Initially, people laughed at Holly's idea. 'You'll never get farmers to meditate,' was the general crux of it.

But she did – and with consistently positive results.

Holly has found that just six minutes of meditation a day 'does the job'. In an eight-week trial, with guidance and coaching, 30 farmers reported a 20% decrease in stress, a 20% increase in focus and a 25% improvement in state of mind. Their progress was tracked using a Mindfulness Measurement Index, where 100 statements are scored from 1 to 10 ranging from 'highly agree' to 'highly disagree'.

In another trial, the farmers meditated alone without coaching or guidance. They still reported a 20% improvement in state of mind, an 8–10 % increase in focus and an 8–10% decrease in stress. Of course, it's highly subjective, far from scientific and the farmers were probably keen to show their support for Focussed Farmers. Still, Holly noticed very little change over the eight weeks in her control groups.

'I just think that it is such a useful tool for farmers,' she says, 'and I couldn't see anything else out there for them.'

One downside is that Holly identified a perception of stigma which turned some of her clients into closet meditators who felt embarrassed about sharing the benefits of mindfulness with their friends and peers, for fear of being judged negatively. I quietly make a note of how much this contrasts with my own, urban attitudes towards meditation – something I would proudly endorse to my friends and yes, probably even brag about a little bit.

Let's lay our cards on the table here – lots of people would consider mindfulness a 'woke' pastime (a common misuse of a word which means being 'awake' to social injustices). Either way, I consider myself woke, and I want people to know I'm an adherer to the self-care movement. So, yeah, I'll talk about meditation. And yoga. And my chakras. Ask me anything. For Holly's farmers it can have the complete opposite effect and send them running for the hills.

'Sometimes we don't do things because we worry what other people will think of us,' she says, 'but as time goes on, I've noticed farmers become more confident as their state of mind improves.'

By approaching mindfulness from a business and leadership perspective, with less emphasis on spiritualism, relaxation and wellness, Holly has recognised the needs, even the prejudices, of her target group and decided to work with them – not against them. As a direct result Holly's newly focussed farmers feel part of something that may have appeared alien before. And they have better mental health to show for it. Letting go of negative aspects of the agricultural mindset – whether it's work martyrdom, ignoring mental health or the 'I'm busier than everyone else' attitude which can drive a sense of rural superiority – while continuing to celebrate all that's wonderful about the 'way of life' (praise be) is within the power of every individual who lives and works in our countryside. Particularly farmers.

To those who never miss a milking: please take a day off. Spend time with your family. Go on holiday in someone else's 'working landscape' – and open your heart to those who wish to enjoy yours. Changing how we think about things is like brushing our teeth with the other hand – it's about breaking the habit.

Sadly, there are many other grave and serious challenges, from poverty to inequality, that contribute to poor mental health and social and economic disadvantage among some of those living the rural way of life.

I say 'some' because it's important to recognise that there isn't just one rural community. It is not a homogenised group, though it may be easier to lump them all together because they look, from the outside at least, far less diverse than our urban populations. A road trip through rural England would give you a pretty firm impression of white, middle-class privilege – and you wouldn't be wrong. There is a lot of wealth in the countryside. We should be wary of swallowing too readily the narrative of downtrodden rural areas as disadvantaged in comparison to those lucky urban areas, with all their buses and regular bin collections. It's far more complex than that.

Dr Ruth McAreavey is a Reader in Sociology at Newcastle

University. Her research focuses on rural development and inequalities faced by migrants in the labour market in regional and rural areas. Before Covid put the brakes on it, she was working on a Defra-funded project looking at 'the lived experience in rural England', which studies eight different communities. In the North East, they selected Barnard Castle (long before Dominic Cummings decided to drive there to check his eyesight during lockdown).

'There are different communities in Barnard Castle,' says Ruth. 'There are the wealthier people who have retired into that community from more urban centres, maybe Newcastle or Durham, and they tend to pick up and start running volunteering schemes and initiatives, at quite a high level, to do with arts, culture and all sorts. Then there are local people, born and bred there, who relate more to Teesdale as their identity. They identify much more with the farming community, so even though Barnard Castle is the market town there is continuum with the hinterland. That's a different community again to the ex-mining and coal areas. There, it is often very socially conservative, people are not necessarily open to newcomers or to wider social change. There are so many complex issues around class, identity and income in rural areas.'

When we talk about 'rural communities', who are we referring to? Is it Lord Barnard with his 55,000 acres or a low-income family on a less affluent estate in East Durham? They both count as rural in the statistics. You could argue it's the same in urban areas – Grenfell Tower burned in Kensington and Chelsea, one of London's wealthiest boroughs, where rich and poor live next door to one another. The difference comes down to numbers. Urban poverty is more visible, and therefore much easier to quantify, for the sad, simple fact there's so much of it and it's usually clustered together.

'The way poverty manifests itself in a rural community is quite different compared to an urban area because of space,' Ruth explains. 'It's not going to be as concentrated, so it's often masked in the statistics. If you look at indices of deprivation, they are all in urban centres. Poverty in rural Northumberland or County Durham pales into insignificance if you look at statistics for the West End of Newcastle where there is dire poverty, and I'm sure it's the same for Bristol.'

To understand rural poverty, we must first abandon stereotypical ideas of what we think poverty 'should' look like. More often than not, mainstream media and popular culture has reinforced the image as homeless people in shop doorways, rundown estates, shoddy tower blocks punctuating city skylines, gangs of youths, drugs, crime, or streets strewn with rubbish.

Rural poverty isn't always visible. It's easy to miss. It can even be disguised as a life you envy and want for yourself. A farmer living in a beautiful stone cottage overlooking the fells – except his cupboards are bare and he can't remember a time when he wasn't lonely. A family out shopping in a pretty postcard village – except they live squashed into a tiny box room or sleep on a friend's sofa. A single mum walking her toddler to playgroup in the quaint village hall – except she can't drive, there are no buses and playgroup is the only place to go; the only thing that breaks the monotony of her isolation.

How to make the public and politicians see such hidden problems? Private suffering buried beneath a beautiful surface?

The contrast between outer appearances and inner turmoil was brought home to me by a third-generation tenant farmer who I met through my day job. He looks after a 350-acre hill farm, and runs 500 sheep, including his beloved flock of prize-winning North Country Cheviots, and a small herd of suckler cattle.

He reminded me so much of the farmers I grew up with on the Welsh Borders, and even my own father, that I felt I knew him instantly. It's hard to pinpoint what they all share – a kind of hard-edged pessimism about pretty much everything in the world, combined with disarming sentimentality. They're rough, tough 'individualists', to use Wade Dooley's description, but when they open up about their land, their animals and way of life they articulate themselves with unparalleled eloquence.

He described his flock as the 'the prettiest sheep in the world,' and chuckled cheekily when he admitted: 'They're as an important part of my life as my wife and children are. And my wife knows that as well!'

On the surface this farmer's life looks peachy. He lives in a beautiful old farmhouse with jaw-dropping views across a national park and he's a happy family man doing a job he loves. But spending time with him over an afternoon, a darker side of his story emerged:

'I'm a farmer who's considered using a food bank,' he said bluntly.

He was leaning on a dry-stone wall in the farmyard, stroking his dog as chickens pecked at the ground around his feet.

'How ridiculous is that?' he asked me. He talked about the guilt he felt buying a £3 chicken from the supermarket, knowing full well the pressure it puts on a fellow farmer to produce it for that price.

'There is a cheap food culture,' he said. 'I'm a victim of it but I'm also a perpetrator of it. I almost unfeather my own nest! I will buy cheap food because that's what we can afford to buy, yet I know that it should be more expensive. But I'm not in a position to pay for it because I don't get paid for it in the first place, and it's a stupid vicious circle.'

He shook his head in frustration and looked at the ground. I suddenly noticed how thin he was; the outline of bony shoulders under a holey jumper; his unshaven face. I could see the strain now, where perhaps I couldn't only a few hours earlier, when I was swept up in the charm of his hill-farming life.

I have never heard anyone sum up the complex problems in our broken food system, the volatility in agriculture and the resulting impact on low-income farming families better. No academic or food justice campaigner or politician could make me see it more clearly. In a few sentences a tenant farmer had captured it all.

The perceived rose-tinted urban view of farming and rural life is a source of endless frustration to those who live and work on the land and it's the precise reason BBC One's flagship rural affairs programme *Countryfile* earned its nickname 'Towniefile'. There are many people in the countryside who truly, vehemently despise *Countryfile*. I know because they tell me. At every opportunity. In their eyes it pedals a fluffy, 'escape to the country and buy some alpacas' message and ignores the less photogenic aspects of country life. What they ignore is the fact Sunday teatime family viewing, the warm-up to *Call the Midwife*, needs to look nice and pretty. It ain't *Newsnight*.

I declare an interest – I worked on the show for many years, still do occasionally – and I genuinely reject the accusation it doesn't deal with the serious, specialist stuff. I'm probably one of the few producer/directors who's made primetime, popular-factual television about herbicide resistant blackgrass and the EU's animal feed protein deficit. These are niche topics that found a mainstream platform, thanks to *Countryfile*. It does not shy away from the darker issues either – rural poverty, depression, domestic abuse, farmer suicides, modern slavery and the inequities of our food supply chain. I also gently point out to its hopping mad critics that *Countryfile* was the canary in the mine – the experiment to see if primetime audiences cared about rural issues.

Turns out they did. People love watching Cotswolds farmer Adam Henson TB testing his cows. The show rocketed from two million viewers in its sleepy old Sunday morning slot to nine million viewers at its primetime peak. In my view a hell of a lot more people know a little something more about our countryside thanks to *Countryfile*. It also sparked a trend, paving the way for programmes like *This Farming Life* and *Our Yorkshire Farm*. Surely that's a good thing.

But farmers would generally choose Clarkson over Craven.

According to the Cumbrian shepherd and author James Rebanks, *Clarkson's Farm* did more for British agriculture in one series than *Countryfile* did in 30 years. He reckons our logic at the BBC is that farming is for a 'niche group of idiots'. Ouch. That stung a bit. And he's wrong. Would a farmer's daughter choose to spend 12 years of her life working with people who thought like that?

He's right though that Jeremy Clarkson has increased awareness of everyday farming life. My friend Stephen Thompson, a pig farmer on the Yorkshire/Derbyshire border, gave 48 media interviews in 14 days when thousands of pigs got stuck on British farms due to the shortage of butchers in abattoirs in October 2021. He told me about a cameraman who, while setting up a shot, said: 'Just stand in that tramline please.'

'You know what a tramline is?'

'I didn't before I watched *Clarkson's Farm*.'

And it did a great job of shining a light on ordinary rural people. I know Calebs, Charlies and Geralds from my own farming community and I've long pondered how to get their unseen faces and unheard voices on mainstream telly. I've tried pitching their stories hundreds of times, to no avail. Turns out you need to be a massive celebrity, with commissioning contacts at Amazon Prime and own a farm you can afford to make mistakes on.

That word 'Towniefile' says a lot about the urban/rural divide. It sums up how a community can feel unseen, unheard, and misrepresented. I'm frequently taken aback by the level of deep-rooted resentment I encounter towards metropolitanism, whether it's aimed at the media and high-profile television presenters, politicians in Westminster or the so-called 'urban masses'.

It's visceral. It's palpable. It's intimidating sometimes. I once gave a talk to a farmers' discussion group in Mid Wales. One of the older farmers marched right up to me, physically trembling with fury, and held his index finger inches away from my face: 'Chris Packham,' he growled.

This is a semi-regular occurrence – perhaps not the finger-in-the-face – but certainly being on the receiving end of pent-up rural rage, and not only from farmers – all sorts of people. I don't take it personally because it's not aimed at me, or anyone in particular (except maybe Chris Packham). It's frustration. The need to be heard by someone outside your community, who might be able to fight your corner in a place where you don't feel you have a voice. In the offices of the BBC, in the corridors of Westminster or in the vegan cafés of London and Bristol.

But the urban/rural divide works both ways. Perhaps even less acknowledged than the frustrations of rural people, are the experiences of urban visitors to a countryside that isn't always warm and welcoming. I experienced it myself during the Coronavirus lockdowns. I got yelled at and told to 'fuck off out of our area' while riding my bicycle (taking my daily permitted exercise) in a semi-rural area on the edge of Bristol where I live. On a separate occasion, after restrictions had eased considerably, I was reminded to 'keep your distance here' while walking through Chipping Campden in the Cotswolds. The pandemic brought out an ugly 'locals only' attitude across rural Britain. It felt strange to be on the receiving end of it. For the first time in my life, I was perceived as a dangerous outsider – and it felt horrible.

Of course, I understand the reasons why. I was also shouting at the TV in frustration at those images of overcrowded beaches and swarms of people descending on remote tourist destinations at the height of the pandemic. In my parent's community, the single-track road to Pistyll Rhaeadr, the highest waterfall in Wales, was blocked with cars when 3,000 visitors a day packed into a tiny village that doesn't even have a GP surgery.

But in the deathly quiet of hard lockdown, rural areas were returned to a time before second homes and holiday lets. My Mum, for one, loved every minute of it and I'm sure many others relished the opportunity to shut up shop and hang a 'closed' sign on the countryside.

Even in my lifetime, it was possible to live an isolated island life in rural Britain. We certainly did. Mum and Dad knew most of the cars, Land Rovers and tractors rumbling up and down 'our' road – they would even toot hello on their way past. We knew who lived in every house and everyone knew who we were too.

This is no longer the case. Most of the houses around our farm are now either holiday lets or second homes. 'Our' road is busier, and the cars drive by faster. I don't recognise many people when I go home for the weekend, and no one has a clue who I am either.

My parents lament these changes, and in some ways, so do I. But change has also brought progress and investment to our area. Old farmhouses and cottages, once crumbling and neglected, are being renovated and restored – at eye-watering costs to the new owners. 'Incomer' money has saved buildings that have been part of our community for hundreds of years. They stand. They survive. You only have to look at the overgrown ruins of abandoned farmworker cottages and barns, marooned in the middle of fields or on the edge of hillsides, to witness a much sadder fate. I've seen these architectural ghosts all over the UK. Once the homes of local families; now all that remains are piles of stone and rusty fireplaces overgrown with weeds. Their 'way of life' lost forever.

Far from highlighting our division, I believe Covid-19 demonstrated how intertwined urban and rural life has become, and how much we need and depend on one another. Tourism is worth £20 billion to the rural economy – without urban visitors, rural livelihoods die. Farmers are key workers – without them supermarket shelves go empty. In lockdown, the nation rediscovered the joy of food. We baked, we cooked and more of us bought local produce – proactively seeking out the farm shops and growers on our doorstep and paying good money for what they produce. Millions of people in towns and cities have learned how to work from home, sparking an epiphanous moment of collective realisation that we can work from anywhere, with many turning their gaze to rural areas.

Everyone's 'way of life' is changing. The relationship between urban and rural is about to get a lot closer and whether that deepens divisions or builds bridges depends on how well we work together.

CHAPTER THREE

POLITICS

I HAVE A HAPPY CHILDHOOD MEMORY of going to the Conservative Club in Oswestry for Sunday lunch with my Nan and Grandad. It only happened once, as a special treat when I was just 11 or 12, and one vivid image stays with me: they served my orange juice in a wine glass. I was utterly delighted. I felt so grown-up and sophisticated, sitting opposite my beloved grandparents dressed in our Sunday best.

I never thought about the political connotations, that Bill and Beryl Jones were obviously Conservative Party members. I don't remember them talking about politics, but then I had more important things to think about – like looking at myself in the window holding a wine glass.

As it turns out, my whole family voted Conservative. Both sets of grandparents and my parents. It's no great shocker as far as the men go – Grandad Bill was a farmer and landowner, Grandad Wilfred was a butcher and small business owner, Dad is a farmer and landowner. Classic Tory voters. The women are more of a mixed bag. Nan Beryl was an NHS nurse and farmer's wife but what made her 'true blue' was an endearing desire to climb the social ladder. We secretly called her Hyacinth. Nan Netta worked in the butcher's shop making pies and didn't give two hoots about politics so probably just copied Grandad for an easy life. My Mum is the surprising one. To meet her you would swear she's a Labour voter. Socially minded and fiercely proud of her working-class roots, coming from a long line of agricultural labourers, she takes a dim view of wealth and privilege. She's a natural socialist, yet she's always voted Conservative. Why?

'I didn't really have a strong opinion,' she says. 'I always thought, whatever happens, nothing's going to make a difference to my little, tiny life. Going to work in a dead-end job from the age of 14. I just wasn't interested.'

Mum left school before her fifteenth birthday without any qualifications, despite showing academic potential and an extraordinary talent for English and creative writing. She wanted to continue her education at the technical college in Oswestry, but it wasn't an option.

'I asked if I could go, but Dad said it was a waste of time educating a girl.' "There's no point," he said, "you'll just end up marrying some 'owd lad! I've kept you long enough – get out there and get earning!"'

Growing up in a rural area, surrounded by Conservative values – it was virtually impossible, unthinkable even, for Mum to break away from the dominant political culture. Plus, she finds party politics utterly boring and switches off immediately if conversation drifts towards Westminster, though I often wonder if this has more to do with her latent left-wing instincts being swiftly snuffed out by her Commie-fearing father.

But she even stuck with the Tories when she got married and left home.

'Your Dad could always put forward a good argument for the Conservatives,' she says. 'And the only person I knew who voted Labour was Pete, and he's a townie. I'm not a townie, I'm a country girl. I don't know – it all got mixed up in my head.'

Uncle Pete grew up on the Caia Park council estate in Wrexham, an industrial ex-mining and steel town in North-East Wales. The estate, the third largest in Wales, has a reputation for high levels of crime and social deprivation, and used to be called Queen's Park before changing to the Welsh name as part of a positive rebrand after it featured in a 1969 sociology book called *The Making of a Criminal*.

Pete went down the pit after he left school at 15 and joined the National Union of Miners. At 17, he signed up for the Merchant Navy and sailed from Liverpool on freight ships all over the world, including Ceramic, a refrigerated cargo ship which brought thousands of tonnes of New Zealand lamb into the UK.

'Dad would have happily sunk that one,' I joke blackly.

Pete gave up his seafaring career when he fell for Dad's younger sister, Gill, in 1973. They met when she was training to be a nurse at the Wrexham Maelor Hospital and spending months away at sea suddenly lost its appeal to a lovestruck Pete. He got a job in construction instead, as a steel erector, but work was contract-based and unreliable, and he was frequently between jobs. He received a frosty reception from Nanny, my formidable great-grandmother, when he visited Craigllwyn Farm for the first time.

'He's unemployed, he's Labour and he's Catholic,' she'd scowl.

Nan Beryl and Grandad Bill were much more welcoming 'at least on the surface,' says my auntie Gill.

'Everything about Pete was alien to my family,' she remembers. 'I know they worried because he was in and out of work and he was from a council house.'

But it didn't faze Pete. He even managed to get on with Nanny as the years passed: 'We used to have a bit of banter, especially if there was an election coming up,' he says. 'I hated Thatcher. I remember there was a sign in one of the toilets at the Stanlow Oil Refinery in Ellesmere Port. It said: There was light at the end of the tunnel, but Margaret Thatcher's turned it off.'

I grew up knowing how my family voted in the ballot box, but that's about as far as our political discourse went. There were no debates, no dinnertime discussions. My blue Dad and red Uncle's political rivalry went about as far as some light jibing about subsidy-grabbing farmers and 'that Labour lot' on strike again. Party politics really wasn't a big deal, and certainly not worth arguing about. Far more influential on shaping me as a person were the 'small c' conservative values I was brought up with, most notably a deep respect for authority: teachers, doctors, the police, the Church of England, Armed Forces, and the Royal Family. We went to Sunday school, gave church readings at Easter, never missed the Remembrance Day service, and always wore a poppy. We are monarchists. Dad gets a bit misty-eyed during the Queen's speech every Christmas Day and I wept watching the Duke of Edinburgh's funeral. I got one detention in school for having my shirt untucked – I was mortified. Rebelliousness is not in my DNA. I am obedient to a fault.

I'm certain there is a class element to this. I have an early memory of standing on the road outside our farm and curtseying to the local hunt as they rode past in their scarlet red coats. I'd seen it on one of Mum's Catherine Cookson videos and obviously cast myself as the peasant in the scenario. The Master of the Hunt smiled and doffed his hat. I never gave it a second thought until I learnt about Benjamin Disraeli in A-Level History; the nineteenth-century Tory Prime Minister who famously said the working classes are instinctively conservative: 'None are so interested in maintaining the institutions of the country as the working classes.' Pride, patriotism, and deference – I recognised my own family straight away.

In terms of character, though, the Joneses are anything but conservative. As kids, we were loud, theatrical, boisterous, and untidy – our house a cluttered sanctum of open books, kicked-off shoes, drying clothes, half-finished homework and old copies of the *Shropshire Star* and *Farmers Guardian* piled high against the wall; a home filled with raucous laughter and Carry On films. We were brought up to be curious about the world around us, interested in people and their stories, and to help those who needed it. We were encouraged to follow our dreams and be ambitious, but also to share what we have, and not place too much importance on money and material wealth. Mum and Dad instilled in us that the best things in life are free, and those things that do cost money? Well, second-hand will do just fine. In the absence of fancy toys and games, Mum taught us to use the power of our imaginations. My sisters and I would spend hours perched on the bouncy branch of the old oak tree, pretending we were trotting along on a horse.

This is the political and cultural backdrop to my childhood. Alternative in some ways, and old-fashioned in others. In terms of a political identity, I didn't have one. Most people find theirs at university. I didn't. My student days were uncharacteristically apolitical. I skipped off to the University of Central Lancashire in 1999, those calm, halcyon days before 9/11 and the Iraq War, in the early years of the Blair Government without a care in the world. As much as Dad disliked

New Labour and seemed to hold Tony Blair personally accountable for the economic downturn in farming, BSE and the Foot and Mouth outbreak, the Government paid all my tuition fees, a lifeline which made a university degree possible for the first time in our house. Of my extended family – aunties, uncles and 18 first cousins – only one other had gone to university before me.

After graduating with a BA Hons in Journalism in 2002, I found a job pretty much straight away as a junior feature writer on a series of posh county magazines, where I covered hunt balls and champagne receptions, and interviewed rich people about their houses. A year later I went the other way, badgering the police, chasing ambulances, knocking on doors, and covering court cases as a news reporter on the left-leaning Wrexham *Evening Leader* in North Wales. Next, I moved on to the features desk at the Wolverhampton-based *Express and Star*, the biggest-selling regional evening newspaper in Britain – though I was tucked away in the sleepy Stafford office.

Life was good, employment opportunities were good, local newspapers were only at the beginning of their depressing downward spiral and I was living affordably in a little rented house in a Shropshire market town near my family. As a graduate and young professional, the Labour Government had looked after me very well. But I didn't vote for them. Dad was having a terrible time – barely scratching a living from the farm, still reeling from the financial and emotional shock of Foot and Mouth Disease and feeling pushed out and forgotten by what he perceived as a metrocentric culture sweeping Britain, epitomised by the likes of Tony and Cherie and their clever London friends who, he believed, saw the countryside as a playground for city dwellers and were conspiring to get rid of farmers altogether. He yearned for a more sympathetic Conservative Government: 'Our party'. But I didn't vote for them either. Worse – I didn't vote for anyone. Mum and Dad's struggle cancelled out my success, resulting in years of political apathy.

I was also in my early twenties and having a lot of fun. It would be disingenuous to pretend I was sitting around, philosophically pondering my political allegiances. I spent most of my free time watching obscure indie bands at the Little Civic in Wolverhampton, or at various house parties in Shropshire dancing to Franz Ferdinand and The Killers.

In 2006, everything changed. I moved to Birmingham and started working for the BBC. Twenty-five years old and living in a city for the first time.

I moved into a large, shared house on Rotton Park Road, where Edgbaston meets Winson Green, and immediately fell in love with Brum. I was blown away by the sheer scale of it and the endless options for things to do – live music every night of the week, theatre, comedy, art galleries, museums, pubs, clubs.

I found drum and bass and discovered the unparalleled joy of dancing all-night in underground clubs and warehouses. My clothes changed, my music changed, I changed. A literal illustration? I bought walking trousers and hiking shoes for my first day on *Countryfile* – I just assumed the office would be full of outdoorsy mountaineering types. Within a couple of months, I was wearing hoodies, ripped jeans and Nike trainers like most other young people working in the TV industry. My eyes had been opened to all sorts of new stuff and only the city could satisfy this new, visceral craving for art, music, culture, and social liberalism. I wanted to push all the boundaries I'd ever known. I had the time of my life.

I'd never lived in a multicultural area before and it was thrilling to me to hear so many different accents and languages, to breathe in the smells drifting from the restaurants and takeaways lining the Dudley Road – shish kebab, jerk chicken and curried goat, pizza, fried chicken and battered cod from our local chippy The Frying Pan. In the general store at the bottom of our street, run by a Jamaican family if I remember rightly, I learned that the fruit that looked like a banana was in fact plantain. Our house consumed vast quantities of fresh coriander, which were sold in big bags for a pound, from the Pakistani greengrocer down the road. It filled our kitchen with a fresh, herby fragrance. Birmingham awoke my senses: it was where I tasted sushi for the first time, knocked back oysters with lemon and tabasco sauce at a fish stall in the Bull Ring Indoor Market and learnt how to use chopsticks in China Town. In the final days before the smoking ban came into force, in the summer of 2007, we enjoyed our last fruit-flavoured sheesha pipe (sometimes referred to as narguile). We liked the apple flavour best, which would be served to our table after a brilliant dinner of lamb tagine at our favourite Moroccan restaurant, hidden away under a car park in the now demolished Paradise Circus. I was out every night of the week, and never seemed to get tired. I could not get enough of this amazing city.

I rode the number 87 bus into work every day, making a beeline for the front seat on the top deck, the best spot for gazing out the window, daydreaming and people watching. The bus would rumble and sway up Dudley Road, inching its way with squeaky brakes past City Hospital. I'd watch the city change as little family-owned shops and scruffy derelict buildings merged into the shiny Jewellery Quarter, new offices and apartment blocks rising higher the closer we edged towards the business district. I'd jump off at Colmore Row, cut through Cathedral Square, along New Street towards the Mailbox, up the escalators by Malmaison and Emporio Armani, past the always empty designer clothes shops, and into the BBC, through the open plan office and to my desk on *Countryfile*.

Those first few years in Birmingham were some of the happiest of my life. It represented, for me, a cultural awakening – discovering a whole new way of living life. A big, urban adventure.

The five housemates I lived with on Rotton Park Road became my best friends, and we remain close to this day. They are like family. Politically though, it was a confusing time, and sometimes deeply challenging.

Without exception, my friends are left-wing liberals. They'd debate politics in a social setting, something, until my mid-twenties, I'd only associated with work, when interviewing town and parish councillors, Members of the Senedd, or the local MP. I had never discussed politics in my own time – remember I was totally apathetic and, besides, politics just made people angry. But my new pals got stuck into Thatcherism,

Socialism and Blairite policies over beers in front of the telly. For the first time in my life, I saw how Conservatism was viewed from the other side of the fence; Tories scorned as backward, swivel-eyed numpties, corrupt privileged morons, unfeeling, selfish, and sly. The sentiment was universally shared; total agreement automatically assumed. It gradually dawned on me just how politically naïve I was. With no party-political identity of my own, I struggled to articulate my feelings. I felt the need to defend something deep within me, something which felt under attack, but I couldn't put my finger on what.

Of course, now I know it was the first time my values had been challenged; that rural conservatism running through my veins. And while I had no affinity with the Conservative Party itself, I felt a deep, emotional loyalty to my family, and my community. I pictured Nan Beryl, who I loved with all my heart, sitting across from me, beaming with pride in the Conservative Club: a woman who'd worked so hard for so little all her life, who was kind and generous to her detriment and endured, with unwavering dignity, decades of hardship on a struggling farm. All Nan wanted was recognition, a little bit of status and to feel special for a couple of hours on a Sunday afternoon.

I felt an enormous weight of responsibility to defend those who weren't there to defend themselves – my Nan, my people, my home. I would try and argue with my friends, I honestly can't remember what over, but I vividly recall the frustration – stammering, tongue-tied, angry at my inability to find the right words, to vocalise the feelings within me. Whatever I said came out all wrong, making me sound like a right-wing idiot. I felt impotent and isolated. My sister Kate, who lived with us for a short while after graduating, told me that Max, one of my closest friends, a fellow journalist and a firebrand of a socialist, had raised it with her: 'I think Anna votes Tory. Does she?'

I was hurt. It wasn't just the assumption that bothered me, that any challenge to left-wing ideas automatically translates into rampant Toryism, but that it had to be whispered behind my back, like staging an intervention: 'I'm worried about Anna . . .'

We've never talked about it, not until I started writing this book.

Fifteen years had passed before I finally asked Max why he went to my sister and didn't come to me.

'That was probably cowardice I think,' he says over a Zoom call from Genoa, where he now lives with his Italian wife Elsa. 'I didn't want to lose our friendship and, to me, calling someone a Tory is an insult, so I had to ask someone else. I probably asked Kate because she would know, and I wanted to understand you a bit better. I didn't ask you face-to-face because I didn't want to offend you.'

'But if I did vote Conservative, I wouldn't be offended, would I?'

'No, you wouldn't. But I didn't think it through that far.'

'Why did you want to know?

'I think it was to do with discussions we'd had about farming, maybe EU subsidies, or not having enough wildlife habitat placement; something like that would have planted the idea in my head . . . and why did I want to know?'

Max looks away from the webcam and thinks about it.

'I do judge people,' he says, 'and a lot more so then. Now I have friends who are Tories but, at that point in my life, I would have struggled. A good part of me hopes that if Kate had said, 'Oh yeah she's always voted Tory,' that would have led to discussions with you about it rather than me just writing you off, but I do still tend to dismiss people if I find out very early in the relationship that they're a Tory.'

We look at each other over the webcam. I absorb the gravity of what he's telling me – I could have lost a best friend before he ever really knew me, based on the box I supposedly ticked on a ballot paper.

Politically, I did find those early months in the city upsetting. It put me through the wringer. On a couple of occasions, I snuck up to my room and cried quietly, feeling alone and out in the cold and just like a terrible person. What was wrong with me? Why didn't I fit in? Why was I different? Looking back, only one thing made me other. My friends came from mixed political and class backgrounds, but are all urbanites, born and bred in towns and cities. Max is a proud Black Country boy from Stourbridge. It's a crap phrase but, to me, they were quite literally streetwise. Compared to them I felt like Heidi the mountain girl. I was the first-born child from a family that had been rooted in the same area for generations, the product of a deeply rural farming community with traditional values that were out of place in this metropolitan world. I hadn't travelled much at that point, had no older siblings to pave the way and simply hadn't formed my world view. I was a political ingénue plonked in the middle of inner-city Birmingham.

It feels good to talk to Max after all these years, to explain how I felt with the articulacy of age and hindsight. Just as we move off the topic and start chatting about something else, Max blurts out: 'Can I apologise at this point? I want to officially say I'm sorry. And I'm sorry you felt so isolated. I want to say I wish you could have talked to me about it, but I know exactly why you couldn't because I would not have been sympathetic. I would have been too busy shouting people down and stamping my opinions on them to stop and listen to their opinions. I never even thought about the fact you came from a different background and a different world. That had never even occurred to me until you explained it just now. I just assumed everyone was exactly the same; yeah, different backgrounds – some rural, some urban – but everyone grows up, well, you know . . . the same.'

I was not expecting this, and certainly wasn't fishing for an apology. The intimacy of our conversation transcends the thousand miles between Bristol and northwest Italy, and I feel like Max is in the same room. I'm deeply moved.

I've never really stopped to think about how I developed politically, but now I remember, I wonder how many other rural kids have moved out from the sticks and experienced that same culture shock in the city, but not had the confidence to own why they're feeling that way?

It sparks another thought – how much do our political beliefs boil down to the basic human need to belong? To be accepted by a peer group?

I did change in Birmingham. There was no great political epiphany; city life just gently nudged me from slightly right of centre to slightly left. I barely even noticed it happening. I formed my political identity through a passive process of osmosis, purely influenced by the people I surrounded myself with – my friends, my colleagues at the BBC and the dominant political atmosphere in Birmingham. I had a salaried job, earned a modest income, hung out with lots of creative, arty people and lived a fairly boho lifestyle. I was heavily into the music and arts scene and spent a lot of time at festivals. I didn't know any businesspeople, had nothing to do with private enterprise and never went to expensive bars and restaurants where I'd be more likely to cross paths with wealthier people. Equally I didn't know any manual workers or tradespeople. I knew plenty of mechanics, electricians, builders, and lorry drivers in Shropshire – but not a single one in Birmingham. We were a bunch of white, middle-class, university-educated, young professionals. I'd moved from one political monoculture to another. Just as finding a Labour voter back home was like looking for a needle in a haystack, finding a Conservative voter in my new life was just as unlikely.

Still, I was finally able to develop a political identity I felt comfortable with. I plonked myself in the middle and, from there, I could still see my family's political beliefs, and engage with them constructively, but equally, as a centrist, I had a kinship with my friends on the left. The way I saw it, I could take the best from both worlds and politely ignore the ideological stuff that made me uneasy.

For 10 years I found a harmonious balance between my urban and rural identities. Sure, I sat on the fence – but I liked it there.

But in 2016, a decade after I first moved to the city, everything changed again. The EU referendum threw my two worlds into a conflict that tore me in half.

I'd left Birmingham by then, after the BBC decided to relocate *Countryfile* and a host of other network programmes to Bristol. To this day I still don't understand why, but it took me to a new city, even further away from the farm, and life got a whole lot 'left-er'. To me, Bristol feels intrinsically political – more so than Birmingham and Manchester, where I also lived and worked. Compared to the UK's second and third largest cities, which are wonderfully liberal places to live and where I found my own freedom of expression, Bristol seems to proclaim its progressiveness from the rooftops. It's a very self-aware city, the only place I've lived where the name is an adjective: 'Oh, that's so Bristol,' meaning quirky and alternative.

'Bristol always has to be different,' says my boyfriend Alex, 'and I like that because anything could happen here. You never know what's going to pop up next.'

He adores this city. He moved to Bristol in 2007, around the same time I was discovering Birmingham, and embraced everything about it. The street parties and pop-ups; brewery tours and neighbourhood arts trails; comedy in tents and circus on ships; performance art in fire stations; nightclubs in the woods; festivals for just about everything under the sun – balloons, boats and campervans, rum, reggae and daal; the largest graffiti festival in Europe and, once a year, thousands of people, Alex included, dress up as zombies and shuffle around the city. Alex is up for everything Bristol can offer and he opened my eyes to the city's imagination too:

'A lot of people think it's worthy and trying too hard,' he says, thinking of his best friend Matt from Bromsgrove who, whenever he visits Bristol, yearns for a simple pint of Stella – not the dizzying menu of craft ales or 'flat cider that costs six quid' in the hipster bars that Alex frequents. 'He'll look at Bristol's street art, shake his head and say, "Everyone here thinks they're Banksy and you're expected to stand around looking at it. Why does everything have to be so trendy? Where's the ordinary?"'

Alex laughs and shakes his head: 'I don't see it like that. Bristol has a strong sense of community and it's good at getting people to join in with things. It attracts people who like to be a bit different and there's revolutionary thinking here.'

'It feels more like a London borough really,' he adds.

I've never lived in London or spent much time there, so maybe that's why I've never fully got my head around the Bristol fandom thing either. Don't get me wrong, I like the city, but I struggle to reach the same level of almost cult-like appreciation for it. Like London, there's so much merch – mugs, cards and tea towels with 'Bristol' sprawled all over them – and, I've just realised, 'London' is also an adjective – meaning cosmopolitan and a bit cocky. Perhaps the mild amusement and gentle cynicism Bromsgrove Matt and I feel towards Bristol's showy-offy-ness is symptomatic of regional Britain's attitude towards, as Andrew Marr phrased it one Sunday morning on BBC One, 'the great condescension of the metropolis'.

I used to go for an occasional swim at the Lido, a lovely outdoor pool in Clifton, the poshest postcode in Bristol. I got chatting to some people in the sauna once – turned out we were all media types – and one of them asked: 'So, when did you move out of London?' He looked flabbergasted when I said I'd never lived in London, as if wondering how on earth I could have ended up in that sauna, if not by way of Shoreditch?

Bristol is like a mini London – with all the culture, convenience, and liberalism of the capital, just with nicer countryside on the doorstep. Politically though, there's nothing 'mini' about Bristol. This city does not pull its punches. Bristol spirit burns a fiery red. Its force pulled down the slaver Edward Colston from his plinth during a Black Lives Matter march in June 2020 and came to epitomise the #KilltheBill protests in March 2021, when peaceful demonstrations against the Police, Crime, Sentencing and Courts Bill were hijacked by a thuggish minority intent on violence. In my mind, anything that messes with a Bristolian's right to peaceful protest is doomed to backfire – like poking a stick in a giant hornet's nest. Many people here identify as rebels of the left, but I've always observed a kind of politeness to their radicalism.

This hardened during the Kill the Bill protests, the mood of the city darkened beyond anything I've ever known. It seemed a Parliamentary Bill aimed at protecting our 'peaceful' daily lives from 'unnecessary interference' served only to transform friendly Keep Cup-carrying hippies into hardened rebels. Alex felt it had more to do with free-spirited Bristolians being cooped up in a Covid lockdown:

'Bristol is a place where people are used to having a lot of freedom to do what they want,' he said in the week it all kicked off. 'I mean, take the street art for example. Graffiti is accepted here, and people are OK with it. They wouldn't be in lots of other places. Bristol is a culture of doing and suddenly, in lockdown, people are being told they can't do what they want.'

Bristol likes to fight the power. Leftism runs through the fabric of the city. It's in the street art — angry hashtags and clenched fists of solidarity — it's in stickers on car windows, leaflets in coffee shops and signs in neighbours' homes. When I'm walking the dog, I read their windows: 'No Nations, No Borders, No Flags, No Patriots. Fuck the Tories.'

A few doors down: 'Live Animal Markets and Factory Farms Breed Killer Diseases. Help Prevent the Next Outbreak: Go Vegan.'

And amusingly on the gates to our allotment: 'Buck Frexit Innit.'

This is the political backdrop to my urban life.

In 2016 I was living between Bristol and Shropshire. I took a career break from the BBC to focus on my Nuffield Farming Scholarship research, which involved a lot of international travel. When I wasn't overseas, I lived at the farm and visited Alex at weekends. If the M5 was a political spectrum, I was driving from one end to the other. Repeatedly.

I considered voting Leave that year.

I was spending a lot of time talking to farmers and a couple of the younger ones had shared their frustrations with me about the European Union's Common Agricultural Policy — the CAP. They were fed up with direct subsidy payments 'keeping old codgers on the land'; vast sums of money being shelled out simply for owning land. In their mind it was keeping young people out of the industry, making it virtually impossible for them to get a foot on the farming ladder while keeping inefficient and unsustainable farming businesses afloat. Subsidies, as they saw them, were a barrier to agricultural and environmental progress.

It resonated with me, seeing these bright and talented youngsters struggling to make a mark in an industry that desperately needs fresh blood and new thinking. I've always identified as a European, and supported the European Union as an institution, but if a Leave vote could help heal our broken food and farming system, then maybe I could consider giving it my support.

That was around February 2016. By the time June arrived, I was away on a research trip for several weeks in the Upper Midwest of the United States. British politics felt very far away. It rarely made it on to the bulletins I caught on Fox News, which accompanied me every morning over breakfast in the lobbies of budget hotels, and the American newspapers were far too preoccupied with Hillary Clinton and Donald Trump to worry about Brexit.

The US had deepening political divisions of her own. In the months leading up to the election of Donald Trump, and during his presidency, I visited the Midwest three times. It was a fascinating time for a rural affairs journalist to explore this Republican heartland, and I've enjoyed some wonderfully open political conversations with American farmers while sat in the cabs of combines or driving around ranches. My good friend Larry Stomprud, in his cowboy hat and jeans, turned to me once while we were checking on his herd of prime Angus cattle on 4,000 acres of rolling South Dakotan prairie and said: 'Well Anna, I'm guessing you would be one of those Democrats – am I right?'

Larry, a rancher in his seventies, was a reluctant Trump voter. He remembers the Republican presidential candidates in order of his preference: 'As I recall it was Huckabee, then Carson, then Cruz, and then Trump. Rubio might have been in there somewhere too.'

Trump was a last resort, but like many of his fellow Republicans in 2016, their disappointment in the chosen candidate wasn't enough to push them towards the Democrats, and certainly not Hillary. From what I heard on the ground that year, a sad collective sigh of reluctant resignation had as much to do with Trump's victory as the frenzied hollering of flag-waving fanatics. Many of those who voted for him wouldn't be caught dead in a Make America Great Again baseball cap.

On my extensive travels in rural America, I never met a single Trump superfan. Not one. I didn't see a Trump t-shirt or a flag or a hat. I saw one bumper sticker which was unusual enough to stick in my memory.

Yet there's this prevailing image of rural Republicans pumped up on Trump love, running around like rabid lunatics swaddled in stars and stripes. I admit to expecting that myself, maybe even looking for it, but I didn't find it. In my experience, there was as much debate about him in rural areas as anywhere else. In small-town Iowa, I met two friends, both farmers, cheerily divided on politics. Brandon Pickard, a Republican voter, liked Trump; Wade Dooley, a Democrat, 'had no respect' for him:

'I can't say I hate Trump because I've never met the guy, but I have no respect for him because he's too dense to understand global trade.'

I met these guys in the fall of 2018 when the trade war with China was in full swing. Despite the slump in demand for their crops and years of depressed prices, Brandon had faith in the 'strong man'. He believed Trump was doing it for the farmers. Wade disagreed wholeheartedly:

'You piss off your biggest trading partner and they're not gonna want to trade with you! Trump is anti-global trade and that's all American agriculture is. We live and die by our trading system.'

Here are two farmers, who might discuss politics over a quiet beer in a roadside tavern in a sleepy town of 500 people in central Iowa. Who's going to hear the subtleties of their views in New York or Washington DC?

What united rural Americans far more than Trump was their shared view of the metropolitan snobbery emanating from the heavily populated east and west coasts, which so often paints them as ignorant and delusional rednecks.

'They want the crazy person every time,' says Wade. 'Every time there's a tornado in Iowa, the media walks straight through the devastation and finds the least eloquent person they can. And the rest of the town goes, Oh God, not them!'

Wade lifts his hands to his head in despair. As if to prove his point I watched the BBC *News at Ten* on the day of Joe Biden's inauguration in January 2021. There was Bruce Springsteen and Lady Gaga and a host of other celebrities looking proud and deferential and generally sensible and intelligent. To represent Trump's supporters, they chose two Texan idiots shooting guns and shrieking 'Yee-hah!' to demonstrate the lengths they would go to in support of their deposed leader. They decided to leave in the bit where Dumb and Dumber accidentally set a stack of hay bales on fire.

This did not represent the people I met in the Midwest. I caught up with Trump voter Brandon Pickard over Zoom in the days following the storming of the United States Capitol:

'We just look ugly,' he said. 'I'm not proud of where our country is, and it saddens me. I voted for Trump, yes. Was he a good president?'

He takes a few seconds to think about it.

'He started turning the economy around, I fully agree with that. Was he an asshole? Yes.'

Brandon has accepted Joe Biden as his president and, as far as he's concerned, that's that. Move on. We talk about it for all of five minutes – there are more important things in his life.

Back to 2016. On 15 June I was on my way home from the US, waiting for a connecting flight in Toronto when I finally caught up with the UK media. I could not believe my eyes. The polls were predicting the Leave campaign to win, but not for any of the reasons I might have supported them. The worm had turned, and from what I could see, not against the EU institution itself but, somehow, against ordinary people. Fellow humans. Immigrants, refugees, asylum seekers. No one seemed to be talking about the pros and cons of EU membership at all.

What happened to the CAP chat? The debate had soured into hate-filled hysteria, xenophobia, and the kind of tribal identity politics I detest. This was just hours before Nigel Farage unveiled his disgusting Brexit poster – that long line of migrants with the headline: 'BREAKING POINT'.

I no longer recognised my country. Somehow, in the space of a few weeks, it was tearing itself apart. I felt sick. I literally could not finish my beer in the airport. Never had I felt so compelled to do something, to say something. Me – she who seeks sanctuary in neutrality – was now filled with a fiery determination to jump down from the fence and join a team.

On 22 June, the day before the referendum, I joined the Remain campaign and volunteered to hand out leaflets in Mold, a rural town in Flintshire, North-East Wales. It was far too late to make any meaningful difference and strictly against the rules as a BBC member of staff, but I told myself I was on a career break, and not on the payroll, and this was something I just had to do. For what felt like the first time in my life, I had an opinion, and, by God, I was going to share it.

It was a strange day. Most people, whatever side they were on, were friendly and gave me at least a smile, even if they didn't stop for a chat. A tiny, feisty old lady scuttled past laden with shopping bags: 'Don't you worry love, I'm Remain!'

But the majority were Leave – most notably middle-aged and older men. They'd acknowledge me with a quick nod and, without even stopping, gruffly announce: 'I'm out!' before marching off with their apologetic wives in tow, who either smiled or shrugged their shoulders. One or two even winked and whispered: 'I'm in.'

One bloke told me to 'get a job'. Another man, bigoted, Islamophobic and visibly shaking with rage, asked if I wanted to spend the rest of my life wearing a burka and praying to Mecca and stated in a loud bellow, 'because that'll happen if we stay in the EU!'

They were the vile ones, but, thankfully, only a minority. Two people out of the hundreds I met that day. Far more damaging and divisive was the general feeling among many Leave voters that they too were cast as a bunch of knuckle-dragging racists. One of my second-cousins who spent a long time deciding which way to vote and eventually plumped for Leave, was furious in the days following the referendum: 'The media was so angry about the result they immediately went out and interviewed the thickest Leave voters they could find.'

By the end of my one and only day of political campaigning on 22 June 2016 I knew we had lost. I was heartbroken. Not only because of what we were losing in terms of EU membership, but the fault line that had cracked open throughout our nation and had sliced straight through me.

That year, I realised the full extent of the conflict between my two worlds. There wasn't a single Brexit voter among my friends in Bristol and very few vocal Remainers in my extended family, though Mum and Dad were a tiny island of 'In' in a sea of 'Out'.

Dad, like many farmers, was a passionate Remainer from the start. There is no solid statistical evidence whatsoever that proves farmers across the UK were 'overwhelmingly' in favour of leaving the EU. Best estimates suggest the reality mirrored the rest of the country: a fairly even split with Leave just in front. But the 'turkeys voting for Christmas' narrative was just too tempting to resist.

Farmers' feelings varied across sectors and geographical regions but, in our local area, Dad was in the minority. In the livestock markets of Welshpool, Market Drayton and Oswestry, he valiantly stuck up for the EU, insisting: 'The French and German farmers are our allies! Their governments care about them, our bloody lot don't care about us. We need them to fight our corner!' But his protestations landed on deaf ears.

Mum would have voted Leave, but just as she succumbed to years of anti-Labour lobbying from her Conservative father and husband, I'm ashamed to admit, this time, she succumbed to the anti-Leave lobby of her three daughters. I feel guilty about that, but she lets me off the hook:

'I suddenly realised the future is for the youngsters. I was watching them on the telly, talking about all the freedom they had to travel around Europe, and all that's gone now, hasn't it? It's made it harder for them. That's why I went with Remain in the end.'

Outside our home, many other members of the family were decisively Leave. None more so than my Labour-voting uncle – Pete.

My first train journey after the winter 2021 lockdown is to North

Wales. I'm filled with excitement at the prospect of a day trip; the same holiday buzz you get on the way to the airport – just, instead of a flight to Corfu, I'm in Coach B of a Transport for Wales train to Wrexham General.

I'm spending a warm April day in the garden with my uncle Pete and auntie Gill – the first time I've seen them since the start of the pandemic. My heart fills with joy to see their faces. Gill has let her hair go white: it looks pretty, shining silver in the sunshine, and Pete says his hair is thinning, but I can't tell. He turned 70 in lockdown and missed out on the big family party arranged for him in the St Mary's Catholic Club. We talk all day long, inching our chairs around the lawn to keep up with the sun until we're finally engulfed in shade and the early evening spring air starts to nip.

'We used to be Great Britain – respected all over the world,' says Pete when I ask why he voted for Brexit. 'Perhaps I'm missing the British way of life.'

Wrexham is one of the 'Red Wall' seats which fell to the Conservatives in the 2019 general election and Pete is one of thousands of former Labour voters who switched. A working-class lad from Wrexham, who supported Arthur Scargill, voted for an old Etonian Tory.

'Did you ever think you would vote Conservative?'

'Probably not.'

'So . . . what changed?'

I look at his house – a detached new-build with a nice garden. Pete has gone up in the world since the 1970s. 'Maybe you've climbed the class ladder a bit yourself?' I suggest.

'Maybe I have,' he nods and leans back in his chair on the patio. 'I remember saying to someone at Brymbo Steelworks, where I worked as a contractor, that I was going to buy my own house one day.'

Pete has always been aspirational. He's been telling Mum and Dad to do up their house and go on nice holidays for years. He believes in working hard to better yourself (he'd hate me for saying it, but he sounds like a Thatcherite):

'I could take you round Wrexham now and I'll see someone on a walking stick, and I'll know they're on the disability and I'll know there's nothing wrong with them. There's a benefit culture isn't there?'

I know there's a Bristolian version of Uncle Pete out there; many, many hundreds. But they're not in my bubble; not in my Bristol. When I drive up and down the M5 it's like stepping through Philip Pullman-type portals into parallel worlds. I shape-shift between two identities, undergoing a chameleon-like transformation. My language changes, my accent changes, my clothes change. I change. This is necessary because I will not choose. I want to be both, but in truth I'm neither. Not fully anyway. I'm the contrarian in the bubble – weakly attempting to challenge, trying to offer an alternative view, but too scared to stake out my territory too forcefully in case I get chucked out. The political divide between my urban and rural lives has stressed me out a lot.

In 2019 our Bristol West constituency went to Labour's Thangam Debbonaire by a landslide, with one of the biggest majorities in the country. Meanwhile Owen Paterson, North Shropshire's Conservative MP for 24 years, sailed through his seventh general election, increasing his share of the votes (before he resigned over the lobbying scandal in 2021). When we talk of urban/rural polarisation, here it is – summed up by two MPs who represented me and my parents.

There isn't a single political conversation I can join in with wholeheartedly and feel part of the tribe. I am perpetually shuffly and uneasy. Surely, I can't be the only one who feels this way?

In all my years at the BBC, I've never knowingly met a Tory colleague, or a Brexiteer. I'm not saying there aren't any, but you'd need the confidence of Jeremy Clarkson to admit it. I can almost feel the right-wingers rubbing their hands with glee at that – just more ammunition to fling at the leftie-liberal elites at the BBC – but that wouldn't be fair. This has nothing to do with some hidden political agenda in its output (by God, bias is a deadly sin at the Beeb. I've had sleepless nights worrying about getting something wrong and putting it on air. I was trained in a meticulous fact-checking culture, scrutinising every sentence and word for balance and impartiality.

Sure, sometimes the BBC gets it wrong – it's run by humans – but it generally protects truth, accuracy, and fairness like the crown jewels).

No, this is more a comment on workplace culture anywhere in the city. Inherently urban institutions, like the BBC, where lots of urban, university-educated, white, middle-class people work. It's easier to fit in if you conform to the culture. I'm sure it would be quite different if I worked for my uncle's haulage business in Shropshire. I would conform to fit in.

The problem with conforming is it gets us nowhere towards bridging the divide. It only deepens it.

I have great admiration for those who know what they believe and stick their neck out. A few years ago, I sat next to a good friend of mine, Robbie Moore, at a farming conference dinner. He's a farmer's son from Lincolnshire who worked at the time as a rural chartered surveyor in Alnwick, Northumberland. We were catching up on each other's news and he told me he was running to become a county councillor at his local authority at the tender age of 31. 'Congratulations!' I said, visibly impressed. 'Which party?'

He fiddled with his napkin and looked a bit awkward. 'Um . . . Conservative?'

There was a definite question mark hanging in the air, as if he were testing the water, not sure if I'd overturn the table in disgust, throw down my napkin and march off. Instead, I wished him all the best. I was intrigued though. We'd never discussed politics before.

'How do you know you're a Conservative?'

He thought about it and turned in his seat to look directly at me:

'Because I believe everyone should have the chance to succeed. There's a baseline, where we look after everyone, but I believe it's OK to reward those who do well, and who want to achieve more by ensuring they have the opportunities in life to succeed.'

I've never forgotten that conversation and over the next couple of years I watched Robbie's meteoric political rise. In 2019 he won the Conservative nomination to run as MP for Keighley near Bradford.

He took the seat from Labour with a majority of 2,219.

I know there's a Bristolian version of Robbie – a young, enthusiastic Tory – but I've never met him. He's not in my bubble; not in my Bristol.

OUR BRISTOL ALLOTMENT IS NEAR Boiling Wells in St Werburghs, a lovely little corner of the city which feels like a commune and an independent socialist state all of its own. There's a house there with a giant banner which stretches right across the front wall. In bright red capital letters, it says: 'PLEASE VOTE FOR JEREMY CORBYN FOR A CARING SOCIETY, A FAIR ECONOMY, HONEST POLITICS.'

When Boris Johnson won the last general election, the banner remained. When Jeremy Corbyn stepped down as Labour leader and Sir Keir Starmer took over, the banner remained. Throughout the Covid lockdowns, the banner remained. I grew more and more curious. Why is it still there? Who lives in that house? 2020 came and went, a pandemic ripped through our population and yet the four-metre-wide banner endured. By March 2021, I had to find out.

I wrote a card with my name and email address. It took me a few days to pluck up the courage to post it. When I eventually walked up to the house, there were lots of other signs – Extinction Rebellion, Black Lives Matter, Not a Penny More to Serco, I Support the Climate and Ecological Emergency Bill, Save the NHS from Privatisation and, right above the letterbox, I Support My Postal Worker. I knew so many of their opinions and not even their names. I posted my card and hurried off.

Later that day an email popped up in my inbox: 'Corbyn Sign'.

My tummy flipped. I suddenly felt nervous. To me, that sign is a megaphone of forthright self-assuredness and total uncompromising political certainty – all things I do not feel I possess. I pictured angry socialists – maybe even communists – who would shout at me and rant. I felt more than a little cowed and intimidated. I swallowed and opened the message:

'Hi. Warning we're serious Corbyn fans so believe in justice, peace, the NHS, the climate, allotments and ending homelessness.

 We are usually working in the garden or workshop so call in next time you're passing.

 Barb.'

 I knock on the window of the workshop, but Mike Harvey can't hear me over the high-pitched whir of his mechanical saw. His back is turned, head bowed in concentration as clouds of sawdust spray in a rainbow arch from the workbench. I wait for a break in the sawing and knock again. Mike shuts off the machine, whips off his safety goggles and waves a hand in welcome as he strides over to the door.

 'Anna! Hello! Come in – straight up the stairs and we'll head into the garden.'

 He holds open the front door and I hurry past in my face mask. It's the first time I've been in someone else's house for months. Covid lockdown restrictions have only just started to ease, and it feels odd that the first home I'm visiting is that of a complete stranger. 'Thank you . . . and I take it you're a carpenter?'

 'No, an interior designer,' says Mike, following me up the stairs, 'but I do build things too.'

 I can tell the house is one of them – it's beautiful. All wooden floors, exposed beams, and large windows, with a log burner in the centre of the room. There's a lot of stuff everywhere – books, CDs, cushions, crafts. It's all open-plan and I say hello to Mike's wife Barb, who's making cups of tea in the kitchen. A fresh, piney fragrance drifts up from the workshop below. This is a calm and cosy home. I feel relaxed.

 Seventy-four-year-old Mike has lived here for 43 years. He raised five children with his first wife in the adjoining house and a couple of his grown-up kids still live there with families of their own. His ex-wife moved just down the road, and they remain friends. Mike has been with his second wife Barb for 17 years and built this extension on to the family home as a kind of designer granny-annexe for them both.

 We take our tea out on to the decking overlooking a pretty garden with a trickling stream. Rescue chickens scratch and peck at the ground in a large hen house, essential protection from the city's thriving fox population. Without me even having to ask, Mike answers my first question:

'We kept the Jeremy Corbyn sign up because it's a statement of who we are. It's what we stand for.'

I don't want to get dragged into Corbyn's policies or the row over anti-Semitism; I want to dig deeper than that. I'm far more interested in what makes Mike and Barb tick as individuals – their core values, who they are. Barb goes first, explaining how her political ideas developed as a traveller in the 1970s when she saw near penniless hitchhikers helping each other out:

'I believe in fairness,' she says. 'We should just share. You give something to someone and then somewhere along the line that person gives something to someone else, and the good swirls round in a mixing pot.'

Barb isn't from a well-off family. Her mum was a widow who held down three jobs, raised five children by herself, voted Conservative and supported Margaret Thatcher:

'She was the typical hardworking woman who was bettering herself. She believed that if you work hard, you get on in life. You don't ask for handouts. She even went to elocution lessons so she could learn how to speak nicely. She hated my Bristolian accent.'

Barb chuckles at the memory. I immediately think of Nan Beryl in the Conservative Club, and how she used to tell me off for saying 'butty' instead of 'sandwich'.

Life took Barb down a different path. During the 1980s and 1990s, as a single mum in Bristol, certain issues started resonating with her – GM crops, the environment and increasingly having to pay for NHS services: 'The dentist always used to be free,' she says, 'and you never had to pay for glasses. That creeping privatisation really put the fear of God into me, and I thought: without the NHS we're screwed.'

She joined Greenpeace and the Soil Association, put her money in the Triodos Bank and voted for the Green Party.

Of the two, Barb appears to be the most anti-establishment. When she states that 'politicians have all gone to posh schools,' Mike gently pipes up: 'Not all – most of them.'

Barb is undeterred: 'They have made their way through life being lucky and they have no sympathy for people who have got nothing. It's like: on your bike, get yourself a job, you lot are scum, scroungers, benefit street! It's looking down on anyone who's not as good as them. I just thought the two main parties were crap.'

'Until Jeremy came along,' chips in Mike. 'It was an awakening.'

Barbs nods enthusiastically: 'The greed, the me, me, me of the people in power is so horrible but with Jeremy Corbyn I thought: there's a chance here. There was a wave of popularity, they were singing to him at Glastonbury; the young ones were all on board; people were all out there fighting for the climate and I thought this is the tipping point. And when they did him in, I just dropped. I went into a complete depression really and I think I had Covid at the same time. I sank into my bed and thought: I just want to die. I can't see a way out of this. That was the moment we could have saved the world.'

Barb slumps back in her chair, spent. I turn to Mike who hasn't said much at all yet. How would he sum up his political identity?

'All mankind is equal' he says simply. 'My heart is filled with compassion for all those who are not of the established way'

I automatically assume he's a communist, but quickly learn his most fundamental belief comes not from Karl Marx, but Jesus Christ:

'I was brought up in a very Christian family and the church was a great influence on me. It gave me my awareness that love is the only way. That really informs my attitude to people – that we are all equal and we all deserve the same treatment. That was there as a foundational thing long before left-wing politics and John Lennon came along.'

Mike was brought up as a member of the Seventh Day Adventist Church. He tells me they are bit like Baptists but keep Saturday as the Sabbath. He was just a year old when his parents left India after Partition and moved to Bristol.

His father was from a British Army family, who'd been in India for generations, and his mother was a quarter Indian, 'which makes me an eighth Indian,' he says.

His Mum and Dad were not political but voted Conservative because, 'it was the gentle and nice thing to do, and the Labour lot were just a bunch of upstarts with unions and things like that'.

Mike himself used to vote Liberal Democrat and wasn't particularly engaged with politics until late in life. Reading between the lines I'm guessing this may have coincided with meeting Barb – but Mike puts it down to someone else:

'It was really Jeremy Corbyn coming along,' he says. 'Politics didn't enter my life for flipping ages, I didn't have anything to do with politics really. But I did lots of social stuff. My ex-wife and I supported street kids around the world and the garden was used for all sorts of fundraising events. We used to have a kids' club on a Saturday. We didn't see it as political; we just saw it as a need.'

For years Mike opened his home to 'people that didn't fit in other places – the dropouts, the drug addicts and alcoholics,' and set up a house church so they could come in from the cold to share food and share their thoughts. It's a word that keeps popping up throughout our conversation – 'share'. So far, Mike hasn't said anything remotely politically divisive or anything you could really argue with. I've certainly never met anyone, except five-year-olds, who would say sharing is a bad thing. And yet there's a whopping great sign hanging off the side of the house that many Corbyn critics would happily have a go at.

'Yeah, we've heard passers-by laughing at it,' says Mike, 'I just pop my head out of the window and try and talk to people about it.'

They are no strangers to political confrontation and relish a debate, but even they weren't prepared for the level of anti-Corbyn sentiment in the run up to the December 2019 general election when they attempted to campaign on one of their core issues – protecting the NHS from privatisation – outside the Bristol Rovers ground on a match day:

'We thought we'd get all the punters as they came in and give them a leaflet about how the NHS is being sold off,' says Barb. 'But that tribe just wanted to watch football and it was: Fuck off out of our way, don't give me your leaflets, let's stick 'em on the ground and tear them up. We stuck it for about 20 minutes and then walked home because we felt like we'd been in a battle. It was in-your-face and horrible. Even someone in a wheelchair! I thought he would support the NHS but it was, "Fucking Jeremy Corbyn, fucking Diane Abbott, stupid cow." We walked home and as we got nearer to our dear St Werburghs and our Jeremy Corbyn banner; it was like a refuge. We're with people who understand and who are nice.'

In hindsight they know it wasn't their shrewdest political move – don't get between a fan and their footie match – but it was the first time they'd witnessed the impenetrable power of a tribal mentality. And it scared them. I think back to how intimidated I felt when I posted my card through their letterbox. Aren't they part of a tribe too? I ask if they know any Conservative voters. There's a brief pause.

'My sister,' says Barb.

'And my sister,' says Mike. 'But we don't talk about it.'

'There must be Conservative voters somewhere in St Werburghs, but they don't admit to it,' says Barb. 'There are no blue signs anywhere. You drive down to Wells and go through Rees-Mogg land and there's wall-to-wall blue signs everywhere.'

They openly admit to not having much to do with rural communities, or even spending that much time in the countryside, aside from riding their bicycles. I get the distinct feeling they're happiest staying in their bubble. How would it be received locally if someone put a Conservative sign in their window here, I wonder?

Mike laughs heartily, a real Father Christmas chortle: 'I don't think they would! I honestly can't see it happening!'

'We'd watch him,' adds Barb, suspicious at the very thought.

'Yeah, we'd certainly want to chat,' nods Mike. 'What's the deal? Why? Or I would probably start by saying: bloody hell, you're brave.'

Would you be friends with them?

Mikes thinks about it carefully: 'I think I could be friends with a Tory if they came out with a plausible argument for what they're saying. I've never heard one, but that doesn't mean you shouldn't give people the opportunity to come out with it and try.'

My final question as I finish off my tea is: what next? Jeremy Corbyn isn't coming back.

'We'll do what we can do,' shrugs Mike. 'We'll vote Green, and we'll grow vegetables and we'll put our energy into practical local projects. If we can't do it politically then at least we can show people a good way of living. But we won't take our sign down.'

The weirdest thing is, I feel the same buzz when I talk to Mike and Barb the Corbynites as I do when I talk to Robbie the Conservative. The conversation flows, I am full of questions, and I learn. My life feels so much richer for having all of them in it, and I wish everyone could know this kind of political openness and diversity. Sure, the hard part is always feeling a bit on the outside – like you're never truly part of the clique – but I've decided the benefits outweigh that. I like stepping through portals, I love the people I meet and the conversations I have. The things they teach me help me make my own decisions at each election, and each decision might be different from the one before, depending on what's happening in the world, in my community and in my own life. Being an urban/rural hybrid has been lonely at times, but it has also enriched my life in so many ways.

The biggest lesson I have learnt is the divide only exists when you see parties and belief systems and groups. When you break it down to individuals and people, it melts away, making you wonder if it exists at all.

Why is there such scorn and collective eye-rolling when most of rural Britain turns blue and the cities turn red at every general election? Why do liberal urban dwellers allow themselves to get so depressed by what is almost always electoral certainty? A good friend of mine takes to social media to vent his rage, demanding to know, 'Who are you Conservative voters?' Because he doesn't know one.

In my mind we should not be beaten down by our differences. Town and country people live and think differently – and that's OK, so long as we understand why. And we can only do that by asking each other and finding out.

The way I see it we have two choices. We can use the electoral map

– rural conservatism and urban liberalism – to prove our irreconcilable differences and give us yet another excuse to retreat ever further into our depressing echo chambers, deepening the urban/rural divide. Or we can use it to inform and improve our national conversation. To give us insight into each other's values and inject some sensitivity and awareness into political debates and discussions.

This is even more important now we've left the European Unions. As we develop our own environmental and agricultural policies it worries me greatly that political identities will get in the way. Not overtly, but subconsciously. An invisible, unspoken suspicion and social awkwardness: I bet you voted for Boris. I bet you voted for Corbyn.

Why is the narrative so often black and white, when the real world is anything but? People can surprise you.

I woke up unusually early one Friday in December 2021. I lay in bed, wide awake, thinking about all the Christmas shopping I still had to do. I reached for my phone. A notification from the BBC News app popped up: 'Lib Dems earn shock by-election win in North Shropshire.'

What?!

I hurriedly read the news, heart racing. Dumbfounded, I jumped on the family WhatsApp group: 'Wow, wow, WOW! I never thought I would see this day. This is history!'

I genuinely thought I would die an old lady before the Conservatives ever lost North Shropshire. It has been one of their safest seats for 200 years, and generations of my family have helped keep them there – until now.

'It'll do the Conservatives good to lose,' said Dad. 'They've sat in that seat for years and taken it for granted. Now they've had a shock and it'll make them work harder at the next general election.'

These are extraordinary times indeed. In 2021, both my parents – lifelong Tories – gave their vote to Helen Morgan. Their new Liberal Democrat MP.

Is something fundamental beginning to change? Are we entering a new era, where assumptions cannot be so readily made about who rural people are, and what they stand for? Dare I even hope that the toxic division, which has defined our political landscape since Brexit, is coming to an end?

We hold in our hands the most precious and fragile opportunity to protect nature and food security for generations to come, as the nations of the UK develop their own environmental and agricultural policies outside the EU. These delicate seedlings will be nurtured equally by people traditionally on opposite sides – philosophically and politically. Farmers and wealthy landowners in wellies, brown brogues and tweed; environmentalists in walking boots, Gore-Tex and holey jumpers, and politicians and civil servants in suits and ties. It's a crude depiction – I know it's shamelessly stereotypical, but I also know it's real. I see these tribes all the time and I can spot them a mile off.

They are very different types of people, often with different lives, values, and politics, but right now they are aligned and united in their mission more than ever before – to build a better country post-Brexit and post-Covid. It won't last long – it is a window of wonderful opportunity to bring liberal and conservative together, to heal some of the polarisation in our bruised society. Practical, durable, and tough – tweed and Gore-Tex have more in common than you might think.

CHAPTER FOUR

DIVERSITY

I MET A BLACK PERSON FOR THE FIRST TIME when I was about ten years old. Our class took part in an exchange project with a primary school in Liverpool – we spent a day in Toxteth, and they spent a day in Llansilin. My memory is hazy, but two images stick in my mind like contrasting polaroids – one is of leaning against a high metal fence in a playground surrounded by grey buildings; the other is of walking across the fields to our local waterfall accompanied by a group of Black children about my age.

The saddest thing is, I can't remember talking to them. I cannot picture their faces or recall their names or stories. I wonder if I was too shy to strike up conversation? I wonder if our teachers made enough effort to get us chatting and playing together? I am not sure they did. I suspect the gaps in my memory hint at a missed opportunity – and a much-needed one at that.

There were no minority ethnic pupils in my primary school and the only nationalities were Welsh and English. Secondary school was almost exclusively white too. Such was the dearth of diversity, a white boy in our year was nicknamed 'Sanjay' after a character in *EastEnders* merely on the basis that he had black hair.

My childhood was a monoculture of white communities, and I never once questioned it.

I met a couple of Black and Asian people at university and my friendship group grew slightly more diverse when I moved to Birmingham, Manchester, and Bristol, but that inner circle, my closest pals, remained stubbornly white. Not by intent or conscious design – we just drifted into each other's lives like atoms, thrown together by circumstance, finding that magnetic pull of shared commonality to form a little white molecule, our Caucasian bubble. Another one. In a nation of white bubbles.

There's nothing remotely unusual about this, particularly in rural areas. According to the 2011 census, more than 98% of Black people live in urban areas. I live in Bristol's Easton ward, where a third of the local population are from minority ethnic backgrounds. This is a completely different world from where I grew up, in a tiny hamlet where ethnic diversity was approximately: 0%.

I doubt I ever would have noticed the complete absence of multiculturalism from my life, or missed it, or recognised its value, if I hadn't escaped a bit of that white homogenous bubble and moved to the city.

In my early thirties I was in a long-term relationship with a Sri Lankan man. We met on the dancefloor in Manchester and shared many adventures exploring each other's worlds, from our farm on the Welsh Borders to his family's home in central Colombo. We moved to Bristol, rented a flat and fell into the ordinary, happy rhythm of life and work. It also gave me an insight into what it's like living in Britain if you're not British, or white. I got a tiny glimpse behind the blinkers of white privilege. There were the everyday annoyances like mountains of red tape and form filling for what felt like the tiniest things, people not replying to our flat share ads, blatantly because of his foreign-sounding name, and, of course, the language barrier. The frustration he visibly felt when trying to convey a message, or a subtlety of meaning, to people who didn't have time, or couldn't be bothered, to listen. Back then, around 2013, the media was overwhelmingly white. I first noticed it on TV commercials. I never saw 'us', an interracial couple, represented in adverts for washing powder or holidays, and once I'd spotted the lack of diversity, I couldn't stop looking for it: on billboards, in magazines and newspapers, on posters in the doctor's surgery. These are the simple things white British people take for granted – the ease with which we can do stuff, making ourselves heard and understood, and having our faces reflected back at us everywhere we go.

I'm certain he experienced more racism than I registered; there were probably even times when I brushed it off, thinking it was better to rise above it than cause a scene. Sometimes, I'm ashamed to admit, I just couldn't see it. I wanted to think better of people, give them the benefit of the doubt. I'm sure they didn't mean it like that . . .

There was one occasion though, in 2014, when my optimistic, white-privileged goggles were blown clean off; when I witnessed for the first time bare-faced racism. Discrimination so blatant, so unfair, I wanted to scream and shout and make the whole world see what was happening.

It happened in the last place I would have expected it, in an environment where I've always felt at my happiest, surrounded by peace, love, and tolerance, where my inner hippy runs wild and free. We were at a small independent music festival, in a field, in the southwest of England. My ex-boyfriend has given me his blessing to share this story.

It was around 10.30pm in the dance tent – I know because I still have the official complaint email in my sent items, detailing exactly what happened. A very drunk woman in the crowd had complained to the security staff that my then-boyfriend was 'acting dodgy'. They marched over, without a moment's hesitation, and demanded proof that the jacket he was holding belonged to him. The jacket I'd bought him for Christmas; the jacket casually flung over his arm while he watched the DJ and danced with his friends. He was like any other festival-goer, smiling, having a good time and minding his own business. The only thing that made him stand out was his brown skin. He was the only person of colour in a crowded tent of white ravers. A bizarre conversation ensued about producing a receipt for the jacket – which obviously we couldn't do – but he remembered there was exactly £6.50 in the pocket. We tried to explain that the woman had been openly staring at us for some time, that this was clearly racially motivated, and totally unfounded. Tempers flared. No one was listening. One of the younger security guards grew more aggressive, repeating that he'd been 'acting dodgy'. A supervisor got involved. He said he didn't believe the woman and told us to get on with our night. Brush it off, pretend it never happened. Of course, we couldn't. We were devastated – but no one seemed to get it. Not even our friends in their happy festival fog could understand why we were so upset. In that moment, as the music kept playing and the bodies kept dancing, I felt like we were the only two people with eyes in a blind world.

We trudged miserably back to our tent, the campsite still largely empty and quiet. A distant baseline thudded in the darkness as the fun carried on without us. Racism had won. I sobbed in frustration. He, far more battle-hardened to the injustices of the world, just seemed weary. The drunk woman kept on dancing. We went home early the next morning, our weekend ruined. To this day, I have never felt such rage, such impotent fury. The security manager never did respond to my email.

It made it worse for me that it happened in a rural area. I felt guilty and somehow responsible – especially as I'd been the one who pushed it: 'You can be whoever you want to be at a festival! Everyone loves each other! It's such a lovely atmosphere!'

I sincerely hope it did not tarnish his enjoyment of the UK countryside, where we spent many weekends and holidays. My family loved him, and my community welcomed him warmly, though of course he stood out. Old ladies in the village hall cooed with excitement at the stranger in their midst one August Bank Holiday at the annual Llansilin Show and Sheepdog Trials, shyly directing their questions at me: 'Oooh, where did you find him?' The spirit was one of surprise, welcome, and curiosity. I felt it too when I visited his community in Sri Lanka. I suspect we both rather enjoyed the attention. We embraced each other's cultures with open hearts and joyful enthusiasm. He was guided along rows of prizewinning vegetables in the village hall, like a visiting royal, and showed his appreciation by excitedly photographing the marrows.

I'm sure the novelty would have worn off – constantly being asked 'where you're from' and feeling like a celebrity because of your skin colour. I'll never know. In the end – as with so many good relationships that don't work out – we were on different paths and wanted different things out of life, so agreed with much sadness to go our separate ways.

We moved on, and both found happiness again, but I thank city life for the fact we ever met at all. The chances of our paths crossing had I stayed in Llansilin, or Oswestry, were next to zero. That would have been such a huge loss in my life – to have missed out on meeting each other. This is the reality for many remote and rural communities – they miss out on the wonders of multiculturalism, and, in my mind, that's a disadvantage in life.

If I hadn't gone to university, or lived in big cities, or travelled the world, would I be the same person? Would I have all the positive aspects of multicultural life to draw upon when I read about suicide bombers or gang violence in the newspapers, or listen to debates on immigration? Could I so easily separate the news headlines from the community, if I did not see that community every day? In Bristol, I live within a two-minute walk of a Hindu temple, a mosque, a church, several Eastern European supermarkets, a Somali-owned café, a British-Asian newsagent, a Chinese takeaway, an Indian restaurant, and a Gambian jeweller. Inclusive multiculturalism is part of my normal daily life. I don't even think about it. Our views are informed by lived experiences, what you see and do every day.

The Black Lives Matter movement has taught us to acknowledge white privilege, and its blindness to many injustices. There is also urban privilege – an exposure to multiculturalism that many rural people simply do not have. Even if they wanted to drop everything and flock to the nearest city, many rural people simply don't have the money or the means to get there. And that too can cause blindness.

It doesn't take a genius to work out the British countryside is overwhelmingly white – 97.6% to be exact. It's no secret that many people in rural areas voted for Brexit and one of their main concerns was immigration. These facts are real, and true. But these truisms have simultaneously led to next-level assumptions and sweeping generalisations that the countryside is full of racists and bigots.

I set out to write this chapter in defence of rural Britain – I wanted to disprove the stereotype. I wanted to say that, in my experience, village folk and farming communities are not inherently racist, that I've personally seen nothing but kind and welcoming behaviour towards people of colour in the countryside. But that's not wholly true, is it? And what use is it coming from me anyway – a white British woman? Am I really qualified to make such a statement? Far from it.

Sadly, racists and bigots exist everywhere, in urban and rural areas, mouthing off and being offensive, threatening, and even violent. But it's wrong to think they alone inhabit Camp Racist; they just sit at the extreme end of a very broad spectrum. Far more awkward to deal with is the low-level stuff (often referred to as microaggressions); that unconscious bias which creeps into our daily conversations, sometimes without us even noticing. I'm sure we've all done it. I've done it. You don't have to be 'A Racist', or a bad person, to unwittingly say something racist. And it pains me greatly to admit that I hear this subtle, opaque language of prejudice more in rural areas and small-town communities than I do in the city. When it comes to diversity and social inclusion, rural Britain has a big problem. I do not want this to be true. But sadly, it is.

I abandoned any hope of trying to deny it the moment I spoke to Essex vet Navaratnam Partheeban, who goes by the nickname Theeb. He's a clinical farm animal vet and works for a practice that covers the south of England. He'll often pop up on social media in his overalls and wellies in a shed somewhere surrounded by Holsteins. Of the two of us, I feel like the farming fraud – I've missed two lambing seasons on the bounce due to Covid and spend most of my time flouncing around Bristol – yet Theeb says something that shocks me:

'I don't feel like I can just walk into the countryside and feel normal. I still can't be me.'

The British countryside is where Theeb lives with his wife and three children, and it's where he has built a phenomenally successful career. Yet he can't be himself? I probe deeper:

'What are we talking about here? Do you mean going for a hike or walking into a country pub?'

'Both. Everything.'

Theeb is a British man of Sri Lankan and Malaysian heritage, born and raised in largely white areas in Scotland and Yorkshire. He was the only minority ethnic pupil in his school and knew the Lord's Prayer off by heart before he understood anything of his own Hindu religion. He has spent his life filtering his identity to fit in – disguising the fact he's the son of immigrant parents; never feeling proud of his heritage; hiding his religion and cultural practices; dreading writing his name on job application forms or arranging house viewings; and generally tolerating other people's intolerance, even laughing along with racist jokes to avoid attracting attention to himself:

'I don't want to call people out when there's no support,' he says, 'I don't want to be on my own so sometimes I let things fly because it feels safer for me than putting my head above the parapet.'

Everyone has their breaking point and, in 2014, the same year as the 'acting dodgy' festival debacle, Theeb met his:

'I had been qualified for a number of years and I was a confident, senior vet in Somerset when a farmer refused to see me. He needed help with a calving, and he'd see anyone else in the practice, but me. It was the final straw. It just hit me that there was no other reason but the colour of my skin.'

Something snapped in Theeb that day. For the first time in his life, he decided to ask for help – but there was no one to turn to; not a single person in his profession stepped up to support him.

'Nobody knew what to do and nobody wanted to take responsibility for supporting me. I just got passed around. When I called the Royal College of Veterinary Surgeons, they told me it was a British Veterinary Association problem, and when I called the veterinary 24-hour mental health helplines they said, "We feel very sorry for you, but we've never dealt with this in the veterinary profession. Racism has never come up before."'

Only 3% of UK vets are from a Black, Asian or Minority Ethnic (BAME) background – a troublingly all-encompassing term for anyone who isn't white basically.

Of farm vets, it's less than 1% – Theeb reckons there are only three or four in the entire country. The membership organisation he was part of – and paying for – had no official policy on racial discrimination. They'd never needed one. The message was loud and clear: it sucks but what can you do? Best to ignore it and carry on.

For months, Theeb tried to do just that but couldn't escape the nagging feeling that something was deeply wrong – and somebody had to change it.

He started writing blogs and speaking out. He contacted an old university friend, Marissa Robson, another vet from a minority ethnic background who had also experienced workplace discrimination based on her race and gender. Together they formed the British Veterinary Ethnicity and Diversity Society (BVEDS), which supports all marginalised communities working in their sector; and that's a huge group with numerous crossovers.

BAME people are also women, migrant workers, working class, people with disabilities, or identify as LGBTQ+, which stands for lesbian, gay, bisexual, transgender, queer/questioning (meaning someone is still figuring out their gender identity or sexual orientation) and the '+' represents other sexual identities such as intersex, pansexual, and asexual.

It took two years, and a progressive new CEO before the Royal College of Veterinary Surgeons finally woke up and, in Theeb's words, 'sorted itself out'. They put together their own diversity and inclusion policy and even inspired Theeb to push further, beyond the veterinary profession and into agriculture – the least diverse employer in the UK.

'We're lacking so much talent in UK farming,' says Theeb, lamenting the racial and cultural uniformity of an industry that feeds a massively diverse population. In his mind, it's a bit rich asking consumers to 'Back British Farming' when it does not reflect the community it serves:

'Why would I care about supporting farmers if they don't care about me and my issues?' he asks, 'They don't represent me. It's like the police. People don't respect the police until they see themselves in the police.'

And it's not just about the people who are missing from our industry – it hurts those on the inside too. Sinead Fenton, who grows vegetables and edible flowers in East Sussex with her partner Adam, happens to be mixed race. I say 'happens to be' because it's totally irrelevant to her identity as a farmer and grower. It shouldn't even be a 'thing'. Yet Sinead has felt singled out because of her skin colour, as if it's down to her to represent the whole gigantic fight for racial equality in agriculture. She was pushed forward for countless media interviews in the wake of Black Lives Matter:

'It had never been a topic of discussion before but suddenly an interview that was supposed to be about both of us ended up being just about me because "being Black is on trend right now." Someone actually said that to me! I felt like my own farming community were looking at me differently. White growers on Instagram started tagging me in things like: "Here Are 5 Black Gardeners to Follow."'

Sinead, who has nearly 10,000 Insta followers, shakes her head in frustration: 'I don't need that. I've worked hard to build my own following. I don't need a white person with 200 followers to put me in their list of Black voices. It happened so much.'

Adam, who's been sitting quietly on the grass in the shade of the caravan throwing sticks for my hyperactive spaniel Lucy, pulls up the sleeve of his t-shirt to reveal an impressive farmer tan:

'I'm still waiting for: "Here Are 5 Olive-Skinned Farmers to Follow."'

We crack up laughing.

'You are very tanned at the moment,' nods Sinead appreciatively, 'we're about the same.'

And they are – in so many ways. They both grew up in East London and first met in college, though Adam's 'emo mullet', a hairstyle which he describes as 'business at the front, party at the back', ruined any chance of romance between them. They finally got together in their twenties, when he had better hair, and Sinead inspired the boy from Romford to quit the office job he hated and join her growing food instead. Two urbanites, neither from a farming background, are sharing an equal and extraordinary journey, forging their own path in a new world. I completely understand Sinead's frustration when the story goes no further than her skin colour:

'I just want to live a life of joy,' she says. 'But what so often happens is we're kept in a place of pain, and we're asked to talk about how hard everything is. Yeah, it is – but I've said it once and that's all I need to say. I don't need to keep revisiting it. If we're supposed to be uplifting marginalised voices, you can't constantly reduce them to what makes them different. You have to uplift them in their places of joy and just see them ... living.'

We haul ourselves out of the shade, heavy-limbed in the baking heat, and wander over to the raised beds where Sinead hands me an edible clover-like leaf. It's from the 'Good Luck Plant', also known as Oxalis Iron Cross due to the striking burgundy markings at the centre of its leaves. A lemony flavour bursts over my tastebuds – way more intense than I ever thought possible for a flower. Suddenly awake and hungry for more, I nibble the delicate petals of *Allium moly*, or garlic flowers, that pack as much punch as a garlic clove (you only need a few flowers to flavour a whole bowl of pasta). I sniff daisies that smell of pineapple; they're too bitter to eat raw, Sinead tells me, but clever chefs flavour butter with them. All this knowledge is at her fingertips, and she exudes the same passion and pride I've encountered on hundreds of farm walks all over the world. It's why I love talking to farmers, and it knows no colour.

The challenge facing us is obvious – UK agriculture is too white, too male, too straight, and just too ... the same. But attempting to change that is a daunting prospect. Even Theeb, who'd made it his mission to improve diversity in the veterinary profession, didn't know where to start with farming. Why not?

'The problem with farmers is ...' he pauses, thinking of how best to word it. It's one a hell of a cliff hanger. I prepare myself for a bombshell – what's he going to say? That most farmers are so racist, they're beyond help? I breathe a sigh of relief when he lays the blame on the lack of clear regulatory bodies. We can work with that.

'Who's the authority on farmers?' he says, 'Who's the influencer? With vets, it's a profession – you're on a register. With farming, where do you go?'

Theeb started working with the National Farmers' Union and sat down with its president Minette Batters in 2018. From there, doors started to creak open and the whitest industry in Britain began having the diversity conversation in earnest – and meant it. Theeb spoke at the Oxford Farming Conference, he was awarded a Nuffield Farming Scholarship to look at ways of supporting an ethnically diverse agricultural workforce, he worked with the Royal Agricultural University on their student recruitment policy, he's a trustee for the Country Trust, a charity which takes children from marginalised urban communities on to farms, and, thanks to Theeb, all vet students at the University of Nottingham will have to complete diversity training and he's helped with the creation of 13 veterinary scholarships for minority ethnic undergraduates, funded by the UK's biggest vet group. The chain of events, triggered by one ignorant Somerset farmer who refused to see a British Asian vet, is nothing short of remarkable.

'He's led a whole movement!' laughs Theeb, 'I'd kept taking it and taking it and taking it and he just got me. In that moment I thought, 'I can't put up with this anymore. I'm going to change it.' So much has happened based on that one tipping point.'

He has no idea if the farmer knows what he started. I hope he does.

There is still a mountain to climb. Even if we rooted out racism completely; if every farmer did diversity training; if every rural business changed their recruitment strategy and actively sought a more diverse workforce – it doesn't change the fact that most people of colour live in urban areas; the percentages by ethnic group in the 2011 census are staggering: 99.1% of Pakistanis, 98.7% of Bangladeshis and 98.2% of Black Africans, live in towns and cities. There are logistical and cultural barriers preventing these communities from accessing the countryside in any great numbers, often combined with negative perceptions of working in UK agriculture, which all need to change if we are to make our rural areas and our farming industry more diverse.

That change is needed as much on the streets as it is on the farms.

My partner Alex shares an allotment with some friends at St Werburghs City Farm in Bristol, a fantastic charity which manages four outdoor sites in the heart of the city, including 13 acres of allotments, a woodland orchard, a community garden, and a smallholding with pigs, sheep, and goats. They run outdoor education courses, training in horticulture and animal care for adults with learning difficulties and loads of activities for young people and families. Alex spent hundreds of hours on the allotment in lockdown and I have no doubt it saved his mental health, wheeling barrow-loads of woodchip and manure up and down the paths, building raised beds, a firepit, a greenhouse, and nurturing his beloved Squash City. We seemed to eat nothing but butternut squash for weeks. In the serendipitous small world of writing this book, I discover by complete chance that Theeb the vet is also a trustee on the City Farm's board and gives them free veterinary advice, a lasting connection from his time at the practice in Somerset.

'It'd be really interesting for your book actually,' he says. 'The Farm is in an area which is 33% ethnic minority: it's in a hugely diverse area. It's perfect because the Farm has animals, there's lots of growing going on, and it's got allotments, so you'd think that's a great place to connect local people with food. But many visitors and service users are white middle class, and a lot of the staff and volunteers are white middle class. Why aren't local populations using it? Why aren't they coming?'

In 2019, a staff member at the Farm, herself a person of colour with mixed heritage, suggested they should be doing more to increase diversity and inclusion, so an Equity Report was commissioned to find the answers. The project team interviewed members of the local community who live within a two-mile radius of the Farm and discovered what Theeb describes as: 'barriers that prevent people from feeling like they belong on the Farm.'

Some of those barriers are physical, such as poor signage, difficult access, and a 'shut-off' location. The report identified cultural factors including a dominant perception of St Werburghs City Farm as a place for well-off, white, middle-class environmentalists: 'hippies', 'vegans' and 'people that ride bicycles' (in all honesty, that was my first impression too.) Some participants said the popular café, which describes itself as 'hobbit-esque', is too expensive and caters solely to the wealthier 'eco' market. Others felt the Farm has a confusing layout and is difficult to navigate, some expressed fears about getting cold outdoors, while many simply knew nothing about the Farm, due to poor communication and outreach. A simple, central message came across loud and clear: This is not a place for us. We do not belong.

To undertake such an honest review was a brave step by St Werburghs City Farm. For a team that's so focused on community, it was difficult to read they were only serving part of their community.

Since the publication of the Equity Report in October 2020, the Farm has commissioned bespoke diversity training for its staff, is looking at ways the café could offer more affordable options (without compromising on food ethics) and started a Community Advocacy programme, helping 10 people from diverse backgrounds in the local neighbourhood explore the Farm and its many services, in the sincere hope word will spread and new visitors will come, and find a place that's as much for them as it is for Alex and his butternut squash.

The Equity Report makes for fascinating reading, and I suspect it's how many Black, Asian and Minority Ethnic communities view the British countryside too – difficult to access and navigate, exclusionary, too expensive, and very cold and wet.

One thing the report did not find, however, was a lack of interest in growing food. It confirmed the complete opposite – that there is valuable knowledge and experience within the local community. Many first-generation immigrants come from farming backgrounds and probably know more about growing a good crop than some of the allotment owners. In fact, the misguided notion that Black people are not interested in farming is a classic bit of unconscious bias rife among white people. The report picked that up too:

'We interviewed the Farm employees,' says Theeb, 'and they had lots of preconceptions that weren't true, like the local population aren't interested in growing when, actually, a lot were growing in baskets and backyards and would love to grow things in the allotments and get involved. A lot of Farm staff thought people weren't interested in the animals, but local residents said, "Well if I'd known about it, I would have come and seen them." There was just this unconscious bias of, "They're not interested in what we're interested in."'

My personal experience of the Farm is slightly different, mainly because our allotment neighbour is Barry, a Jamaican man in his eighties who's been growing vegetables there since 1980. And there are the younger guys that help him manage his patch: Sam, a Sri Lankan chef and restaurant owner; Conrad, a Jamaican taxi driver and singer; and Joe, a Bristolian musician. We got to know them all in lockdown – Sam cooked incredible feasts on a little gas stove inside an old rabbit hutch and spoiled us rotten with paper plates piled high with rice and daal, and fragrant curries of chicken and lamb. There's a wonderful sense of community there. I'm smiling even as I write this, thinking back to those sunny lockdown days on the allotment.

On a bright but chilly early spring day, Conrad wanders over to our picnic table for a chat. He's a farmer's son, raised on a mixed farm in the northern parish of Saint Ann in Jamaica, which, he's quick to tell me, is close to where Bob Marley was born. Conrad moved to Kingston for a while and worked as a street seller, learning what he describes as the 'skills of survival' in the city, before setting up his own four-acre farm near Old Harbour, on the island's south coast. The growing conditions were different there from in the north, allowing him to produce high-yielding crops of okra, peppers, pak choi, and callaloo, a green leafy vegetable, a bit like spinach, which is popular in Caribbean cooking:

'Those crops, if you understand them, can turn your weekly salary, while potatoes and yam, which grow better in the north, you might have to wait three months or nine months before you can harvest . . .'

He snaps open his sports bottle.

'. . . so, if you understand how to do it well, you can earn a weekly wage from it.'

He takes a sip of his drink – a kind of homemade brew

made with a warming mixture of turmeric, ginger and Irish Moss, a type of seaweed which grows abundantly on the shores of the North Atlantic. Conrad also raised pigs and goats, had three cows, and employed staff. It was a thriving business but, around 2000, he ran into problems with irrigation and water abstraction and his crops failed. 'I was really gutted,' he says, 'but there are different chapters in your life, and I had to be prepared to move on to other things.'

Moving to the UK was not one of them.

'I never had no intention to travel, I was happy farming in Jamaica, but there was a friend of mine, who I used to help out when we were young guys, who had moved to England and he asked if I wanted to come for a holiday. At the time I think, "Just go for a break and come back," so I take up the opportunity.'

Conrad came straight to Bristol and loved it, so applied to extend his stay and do some studying. Twenty-one years later and he's made a new life for himself – working first as a Bristol bus driver and later setting up on his own as the 'taxi man driving the little black taxi van', which he sings about under his artist's name Tru Tryah (check him out on Spotify). He writes and records reggae songs with gospel influences and calls it 'clean and inspirational' music designed to lift people's spirits in difficult times. His songs say a lot about him, and how he sees the world. In 'Share the Food', released in 2021, he begs 'Mr Chef Man' to make sure the whole nation has enough to eat.

'You'll never go without food if you're a farmer,' he says pointedly.

Growing is more than a hobby for Conrad. It's not just a nice thing to do at weekends for a bit of fresh air – he treats the allotment like work. There is security in self-sufficiency, something he has always relied on.

'You see, farming is my trade,' he explains. 'I see farming as the biggest trade on the planet because the farmer feeds everybody.'

He felt disconnected from his trade when he first moved to Bristol. He missed farming. He needed to farm. He made do with 'planting up' the backyards of several different houses he lived in around the city until he met Barry in 2020, who needed an extra pair of hands on the allotment. His eyes light up when he talks about it:

'I have a special love for farming. I have been growing stuff since I was young, from eight-and-a-half years old, when I was with my dad and he take me to the fields. Farming and singing are the most natural things for me to do because I can do them without anyone else's help. I can do them on my own.'

Conrad has gone to huge efforts to hold on to his farming identity – a matter of great pride for him – but not all people from agricultural backgrounds share such positive memories of the land. There are those who leave their countries of heritage to escape the grinding poverty, drudgery, and boredom of a subsistence existence to seek a better life in the city. Most descendants of immigrant farmers become citizens of the cities – the same cultural disconnection that occurred during the Industrial Revolution when vast numbers of British people left the countryside, seeking a better life too.

Many urban communities, Black and white, have been separated from the land for generations. We're back to that collective amnesia I talked about earlier – forgetting the farmer DNA that most humans share. For people of colour, though, there are other barriers which can prevent them from reconnecting meaningfully with the land.

Banseka Kayembe and I share a lot in common. We both grew up in small, mostly white communities in very rural counties – me in Shropshire, Banseka in Suffolk. The thing that divides our childhood experience is that I felt embraced by my community; clutched to its bosom as a nice local girl from a nice local family. Banseka felt like an 'outsider' who 'stuck out like a sore thumb'.

'You will probably always retain a sense of nostalgia for where you come from,' she says to me over Zoom from her parents' home in Felixstowe.

I nod enthusiastically. 'I do, yes.'

'I will never feel that for this place.'

Only one thing makes us outwardly different. I am white. Banseka is Black. I felt like I 'belonged'. She didn't.

Banseka's Indian mother and Congolese father left multicultural North London in 1997 in pursuit of a slower pace of life and a bigger family home than what they could afford in the city. They chose Felixstowe, a sleepy seaside town on the Suffolk coast, to raise their son and daughter.

'We really did feel like the only Black people in the village,' she says.

At just four years old, she already felt 'slightly alien'. I think of the four- and five-year-old children I know – they seem so innocent, carefree, and oblivious. Banseka sees the surprise register on my face:

'This whole idea that kids are innocent and don't understand this stuff is rubbish. They pick up social dynamics very quickly, even if they don't have the language for it yet. I remember my older brother saying that he wished his name was Jason; that he didn't want to be called his Indian name. He was only seven or eight and he'd already absorbed that his name wasn't something to be proud of. We understood that we were different and didn't fit in.'

Banseka has always wanted to leave, but her parents are happy with their chosen life in Felixstowe. Despite its many challenges, they always felt it was 'worth it'.

'Sometimes I disagree with them if I'm honest with you. I can't help but wonder what kind of person I would be if I'd grown up in a city. Would I be different? My plan is to go back to London as soon as possible.'

Banseka is a freelance writer and runs Naked Politics, a media platform dedicated to engaging young people in politics. Like so many 29 year olds, she is living with her parents until she saves enough money to buy a home of her own. She yearns for the anonymity of the city:

'I think a lot of white people associate the countryside with safety, and view the city as a dangerous, urban place, but for many people of colour, it's the reverse. The countryside or white suburbia is a place of extreme discomfort and a place where you can never quite relax or let your guard down, not like I can in London anyway.'

'Is that because the countryside is racist, or because you feel more conspicuous there?' I ask.

'That feeling of insecurity, because you don't look like everyone else, comes from the fact the place is inherently racist. I don't want to put cities on a pedestal as safe havens where nothing bad ever happens – you still get racism in the city – but it's a lot more subtle. Out in the countryside, or by the seaside, it feels more acute and out in the open.'

Banseka has countless examples – like her dad getting continually stopped by the local police in his car, until they finally realised they were stopping the same Black man again and again, who clearly wasn't doing anything wrong. She's heard the 'N' word being casually flung around. Then there's the subtle unconscious bias – people asking if she has a nickname because they'll 'never remember' her full name and make no attempt to even try.

'That wouldn't be acceptable in more urbanised areas,' she says.

Banseka has found a sense of belonging in the city, and I get the feeling she wouldn't miss a thing about a more rural life. She seems a bit bemused when I ask if she enjoys hiking, or cycling, or getting out in the great outdoors:

'Um, like shooting and Barbour jackets and stuff? No, not really. The Lake District is nice, but I guess that's kind of touristy.'

Banseka knows where she wants to be. She has chosen the city over the countryside and feels empowered in that choice. For many communities of colour though, that choice was taken away from them.

A deep trauma associated with the 'fields' endures to this day, particularly in the US. This trauma was sown during the transatlantic slave trade – which spanned 400 years from the fifteenth to the nineteenth centuries – when millions of Africans were kidnapped from their homes, packed into the dark hulls of wooden ships, and transported across the ocean to be sold into slavery; toiling in the fields for white masters. How anyone could expect the trauma of such unimaginable collective human suffering to just disappear with time must be off their rocker. Time does not erase – it buries. Generations of Black families left their farms – or were forced violently off the land – and poured into the cities with no choice but to live in the poorest neighbourhoods, where the ghost of colonialism and slavery faded but forever haunted. It has led to a tragic disconnection from land, food, and nature, contributing to America's overwhelming health crisis, with chronically poor diets among disadvantaged inner-city Black populations.

I confess I had never consciously acknowledged this lasting trauma or considered how the long shadow of slavery was robbing people of their chance to enjoy nature and grow food, until I met a Black farmer on the black dirt soils of Orange County, New York:

'I remember telling one of my closest friends that I was farming, and her response was: we spent so much time trying to get away from that.'

Lorrie Clevenger grew up in rural Missouri, the only person of colour in a small-town community. She moved to New York City as soon as she could, finding blissful anonymity in crowds of colour, but something unexpected happened in the city – she yearned to farm; something deep in her bones was calling her back to the land:

'I started gardening in a community plot in the Bronx and when I started growing my own food and started eating the tomatoes from my own garden there was this sense of freedom. The only way to describe it is as a spiritual experience.'

She says it was 'liberating and healing' to go full circle; from a challenging childhood in rural America, to finding solace and sanctuary in the soil, nurturing something deep within her soul. Lorrie wants to share that feeling by empowering other people of colour to throw off the shackles of a toxic legacy and reclaim a positive relationship with the land:

'The enslavement of African people was severely traumatic and those are the stories we're told, but there's also a lot of history in Black agriculture here in the United States; of Black people who never experienced slavery, who were actually free when they came to this country, or who bought their freedom and started their own farms.'

So many of these positive stories have been lost. In 1920, there were nearly one million Black farmers in the US. Today that number has dwindled to less than 50,000 – just 1.4% of American food producers identify as Black. Lorrie is on a mission to reverse this trend:

'We were growing food in the communities that we came from in Africa, and it was a beautiful relationship that we had with the land and with each other, and that has been lost through the brutal experience of being a people enslaved and forced to do work for other people. The land was the scene of the crime, but it wasn't the perpetrator of the crime.'

Lorrie threw herself into the urban farming and food justice movement and eventually ended up in Orange County as a co-owner of Rise & Root Farm, a three-acre organic vegetable farm leased from the Chester Agricultural Center, a not-for-profit which provides affordable long-term tenancies on prime agricultural land. It's 2018 and I'm interviewing Lorrie for the BBC World Service in the high tunnel greenhouse where heritage tomatoes – red, yellow, purple and orange – grow ripe and plump on tall vines. Struggling with the humidity, we step outside to breathe in the cool October air. I look across the flat fields all around us. The black dirt crumbles through your fingers like compost; many large-scale arable guys would give their right arm to farm these first-grade soils. I've watched an Australian grain grower, used to dry thirsty ground, crouch down in a similar field in northern France, scoop the soil into his hands and swoon: 'Don't you just want to put cream on top and eat it?'

I doubt a plough, seed drill and sprayer will ever work this land again. These 270 acres were taken out of conventional crop production in 2015, carved up into smaller plots and handed over to enterprises like Rise & Root, which all use organic, regenerative and agroecological farming principles. Lorrie believes these are just fancy new words for what indigenous people, 'the ancestors', have been doing forever anyway:

'Agroecology is really referring to indigenous practices of growing food all over the world that people have been doing for centuries without the chemicals, without the large machines. It's not a new idea. It's been around since humans learned the art of cultivating food,' she says.

Lorrie joined forces with three other New Yorkers – fellow urban farmers and community activists Jane Hayes-Hodge, Michaela Hayes-Hodge and Karen Washington. Together they grow vegetables, edible flowers and potted plants for city dwellers and farmers' markets in Manhattan, Brooklyn and the Bronx.

They are a diverse team – a same-sex married couple, two people of colour, and all four of them women. Perhaps that shouldn't be a point worth remarking on in the twenty-first century, but in the context of US and UK agriculture, we're still in eyebrow-raising territory here, and I think it would be a mistake to pretend we're not.

In August 2016, the ultra-conservative radio talk show host Rush Limbaugh, who died in 2021, took issue with a rural pride agricultural conference, supported by the US Department of Agriculture, aimed at opening up agriculture to more people from diverse backgrounds. Limbaugh felt it was a subtle campaign orchestrated by the Obama administration to 'bust up' rural America, the last bastion of true conservativism, by infiltrating it with lesbian farmers.

My first instinct when I heard this was to laugh out loud at the absurdity of it: lesbians as sinister agents of the state, sent to sow the dangerous seeds of liberalism alongside their organic cucumbers, while homophobic country folk barricade themselves inside their houses with pitch forks yelling: 'The Lesbians are coming!'

It's insulting to both rural and LGBTQ+ communities – and all those who identify as members of both. For gay and transgender people seeking to leave the city and start a farming business in the countryside, this kind of toxic rhetoric isn't funny – it's threatening.

When searching for land to set up their business and a place to live, married couple Michaela and Jane were careful to choose a rural community where they felt safe and welcome. They eventually settled on the small town of Chester, less than a two-hour bus ride from Manhattan.

'Our county – though close to NYC – is still very conservative, and we are fairly isolated from the LGBTQ+ community here,' says

Michaela.

To bring them closer, they host monthly community workdays aimed at encouraging LGBTQ+ people, who perhaps have never felt comfortable venturing into the sticks before, to travel out to the farm and experience rural life for themselves:

'We usually work for a couple of hours and then sit and have a meal together and we've had transgender people break out in tears because it's the first time they've ever felt safe in a place outside of the city. To me that's part of the gift we have and it's our responsibility to share it.'

Rise & Root is one of the most inspirational farms I've ever visited – I loved it there – but I don't want to overstate the agency that 'incomers' to the countryside have in this issue. It's not down to city folk to fix the countryside, nor drag it kicking and screaming into a more liberal and tolerant age. Indigenous rural communities are simultaneously on their own journey towards becoming more open and accepting of difference. They may not shout about it as much, and it may be slower, but positive change is already happening.

When I was growing up on the Welsh Borders, I was told categorically that there were no gay people in our community. I used to question the elders on this:

'Maybe they just don't talk about it?'

'Not likely! There are no queers around here!'

Oh. OK then. We live in a hermetically-sealed-homo-free zone. End of conversation.

It makes me sad to think of all those generations of closeted gay people in the countryside, who died without ever sharing their truth, who probably even nodded along and agreed there were 'no queers here'. No one should ever have to endure that kind of loneliness.

Coinciding perfectly with a much more open national conversation, particularly on social media, around mental health, sexuality and gender identity, things have changed quite dramatically in the countryside. Dad buys and sells lambs to a trans farmer and, off the top of my head, I know of three local farmers' sons who have come out. One of them is Gethin Bickerton, the actor in Cardiff. The first person he told was his drama teacher, when he was 17, but it was another couple of years before he was ready to tell his family he was gay:

'We'd never been the kind of family to sit down and talk about emotions; I'm sure like many rural families. I had this irrational fear of people changing their opinion of me, or how they behaved towards me. Why am I having to say this to someone? It doesn't change me in any way. I'm still Gethin.'

He found it easier to tell his sister, a theatre director, and his Mum, who also went to university in Cardiff:

'She had met gay people and saw that they were normal and didn't have three heads!'

But he found it trickier to broach the subject with his dad and brother, 'who maybe only had the livestock market as their connection to other people in the world.'

In the end, television helped pave the way. With greater coverage of LGBTQ+ issues in the mainstream news and more sports personalities speaking publicly about their sexuality, Gethin felt it opened the door:

'The more it was talked about, the more normal it became. My dad was the last person I told but he was the one whose response I cared about most. And he was absolutely brilliant. He just said, "As long as you're alright, that's all that's important." I have never cried so much. I had a positive reaction but I'm so glad I waited for the right time. If I speak to other farmers, through my work at the DPJ Foundation, the one thing I always say is: work it out in your head, take your time, there's no rush and there's no pressure at all. Regardless of the response you get, there will always be people to support you, who are there to help.'

In 2019, there was a rainbow tractor at Brighton Pride. It was the first time British farmers had been represented at a gay pride event and they went down a storm – 50 members of the ag community proudly parading their Sassey Ferguson with #OUTONTHEFARM emblazoned across the window. It was organised by AgRespect, co-founded in 2018 by Lincolnshire farmer Matthew Naylor to promote diversity and inclusion within the farming sector.

Matt grows cut flowers on reclaimed silt land around the Wash in South Lincolnshire. He's one of the largest players in the flower market, producing more than 50 million stems of daffodils, delphiniums, alliums, asters, sunflowers, Sweet Williams, gladioli, and peonies for supermarkets in the UK and mainland Europe. Matt's family have been farming in this area since the 1600s, though he's prone to exaggerating for the benefit of supermarket buyers who love a bit of history: 'I think I'm saying the 1500s now.'

Matt is a character. In some ways he's your quintessential Barbour-wearing, Discovery-driving, well-to-do, landowning country gent with 700 acres who spends a lot of his time mixing with the farming top brass at well-heeled industry events. He's also a socialist (admittedly a champagne one), a Green Party voter, social justice campaigner, and a gay man.

'I'm gammon with a twist,' he says, making me snort with laughter.

Rewind to the 1980s when Matt was a shy teenage farmer's son – deep in the closet, standing around awkwardly at house parties, trying to fancy girls. He grew up in the same farming culture as I did: no queers here! A rural LGBTQ+ network was unimaginable. That he would one day lead it? Unthinkable.

'Being gay wasn't a thing,' he says. 'It didn't even exist. No one declared that as their sexuality and the few examples we did see would always lean back to the real left-wing stuff of the time; Arthur Scargill and all that. Anything progressive was viewed as left-wing and anything left-wing was trying to rip society down. We didn't even know left-wing people, let alone gay people.'

Matt finally came out to his friends and close family in his twenties. He was in his thirties before he felt confident enough to be fully open about his sexuality and live an authentic life, as his true self. It turned out to be the best thing he ever did.

'I'm so much better at my job now. Being different is hugely valuable when trying to differentiate yourself in a crowded marketplace.'

He freely admits that the power and privilege that comes with being a boss of your own company provides some protection against homophobia – people are more likely to show you respect when you're paying their wages, and it's probably made life a bit easier for him. I ask Matt if 'being gay' is more difficult in a rural area compared to an urban one. He seems amused by the question and points out that 'being gay' is as easy as downloading an app and secretly hooking up with someone, and you can do that anywhere. But finding love in the countryside, a long-term relationship, and meeting someone you can settle down with – that can be a lot harder:

'It's very difficult to meet people because there are fewer people,' he says, almost shrugging it off – after all what can you do about the population maths?

'It's also difficult because you tend to live closer to the people you grew up with. There aren't many, straight or otherwise, who wouldn't feel inhibited by that. No one wants to do their dating under their parents' noses. It's too much scrutiny. So, geography is a big part of it.'

Just as I'm thinking, 'Well, that's the same for everyone, whether you're gay or straight,' Matt seems to read my mind:

'That's only the practical bit. The cultural bit of living your sexual or gender identity authentically is about shame, that's the bit that really traps people and stands in the way. Shame is the biggest barrier. I think that, in rural areas, people have less experience of dealing with these issues and they're not as quick to call stuff out in public. It's more difficult to avoid racist and homophobic and sexist people in the countryside. In the city, if someone made comments like that repeatedly you'd hopefully try and see less of them. But in rural areas that isn't so easy when a racist or a homophobe is actually a really good tyre fitter or is the only person who can supply you with oil filters for the tractor.'

But is it necessarily any easier 'being gay' in the city? I ask my friend Letty, who identifies as queer:

'In a city you can find your people,' she says. 'There are so many more queer spaces opening up now, where we can kiss in front of people, men wear loads of make-up, and people are really flamboyant. It's wonderful! But for people in the countryside, they can't do that. I guess it doesn't exist. They might be able to get together with a couple of friends, have a little party in a house or go to a nightclub somewhere, but it's not going to be this wonderful ability to express yourself.'

Letty grew up in a flat above a butcher's shop in Sidcup in the London Borough of Bexley. Despite its proximity to hugely diverse areas, Letty describes it as a 'samey' place where she felt out of place.

'There was a cultural sameness. The mentality was you grow up, you buy a house, get married, and have children, and I found that very difficult. I didn't accept my sexuality for a very long time.'

Letty found herself searching for 'difference', first by choosing a college closer to Brixton, where the students came from more diverse backgrounds, and then going to university in Brighton. But she was in her mid-twenties and living in Montreal, Canada before she finally came out to her best friend, while sitting in a tent, in a park, watching a 'gammy pigeon' wobbling around.

Now, 10 years later, she is in a long-term relationship with a woman and has moved from our community in East Bristol to live with her partner Abii in Weston-super-Mare. But an urban life doesn't automatically equate to an authentic life as a LGBTQ+ person:

'In Bristol we would hold hands in the street, but we'd still pull away if we weren't sure about the person walking towards us. In Weston we're probably even more cautious. In the evening we would never hold hands because you could be attacked. That's the reality for us. We live our lives as we want in our home but outside, you're continually assessing the situation and working out if it's safe. And that would be the same in Bristol. From that perspective, there's no difference wherever you live.'

Back to Matt. There's a very basic, and difficult, question gnawing at the back of my mind, so I decide to come straight out with it:

'I've heard it said before that rural people are uber conservative, even racist and bigoted, and they want to maintain the status quo and never change. How fair is that assumption?'

Matt is unfazed. In fact, he relishes the question, like he was waiting for it:

'I'll give you hard figures. 75% of the people in the constituency I live in voted for the incumbent MP who's a Tory and a Brexiteer who believes in bringing back the death penalty. He's got one of the safest Tory seats in the UK.'

Matt, who votes for the Green Party, is referring to Sir John Hayes who's been the MP for South Holland and The Deepings in Lincolnshire since 1997 and re-elected seven times.

'So, your answer is...?'

'Of the 25% who don't vote for him we have to assume that some have more extreme views and voted for the Brexit Party, and some were poorly and couldn't get to the polling station, and then that leaves me and one or two others.'

I had hoped for a more nuanced answer, one that didn't lump people together in one camp. But if we're looking at it purely through a parliamentary political lens, then there's no arguing with the fact Conservatism is deeply ingrained in these very rural areas – 'thanks to the landed people and a non-radicalised working class' says Matt, dripping with sarcasm.

I should add that being in favour of capital punishment does not automatically make you a racist or a bigot. Weirdly, I have some experience of the death penalty debate, and it does tend to land well in rural areas. In the mid-1990s, when I was about 14, my parents drove me all the way to Machynlleth for a Young Farmers' debating competition. I had to argue for bringing back capital punishment. Looking back, it must have been a bizarre thing to behold – a short, chubby-faced schoolgirl with big hair thumping the table: 'An eye for an eye! A tooth for a tooth!'

Still, coming from a staunchly Conservative voting family, I think Tory MPs would be wise not to flatter themselves too much that they as individuals, or their policies, have anything to do with them winning big in rural areas. You could slap a Conservative badge on a horse and the true blues would vote for it.

I come from an almost identical parliamentary constituency to Matt's. Until his resignation in 2021, Owen Paterson was our MP in North Shropshire for 24 years. I was in school uniform when he took over from his predecessor John Biffen (who held the seat for 22 years), and just like death penalty man, Mr Paterson was re-elected seven times. Dad can't stand him. He doesn't agree with his hardline Brexiteer politics – he's a Remainer – and finds him 'patronising' – and this was before the lobbying row. Yet Dad still voted for Paterson six times. He strayed once, in 2019 to the Lib Dems, in a half-hearted effort to get some fresh blood into the seat, but it felt like such a betrayal, to the Conservatives and himself, he didn't think he could do it again.

'It was a protest vote really,' says Dad, 'but it was weird. I almost had to get my left hand to pull my right hand across to mark the paper. Loyalty I suppose. It's built in you.'

But he did do it again. In 2021, he dragged his right hand across the ballot paper to vote for Helen Morgan, and North Shropshire turned Liberal Democrat for the first time. Fresh blood at last.

Sometimes, voting Conservative in a rural area has nothing to do with politics – it's more about identity. For that reason, I must respectfully disagree with Matt Naylor's insinuation that a large Conservative majority is a way of measuring prejudice and bigotry within a community. I have more faith in people than that.

I know traditional values run deep in many rural areas, and outdated views can be difficult to root out (capital punishment is not coming back folks). But I do not believe in disregarding well-meaning people because of the world they inhabit – a world devoid of diversity, for instance. Just as there are logistical and cultural barriers preventing people of colour from exploring rural areas, I believe there are logistical and cultural barriers preventing small rural communities from exploring multiculturalism and getting comfortable with diversity. I believe this disadvantage provides the perfect growing conditions for unconscious bias – but admonishing people for it doesn't work. Believe me, I've tried preaching from my high horse, and it gets you nowhere. Shaming leads to humiliation, and humiliation leads to hiding away. Hidden prejudices have already festered for too long and look where that got us.

At the same time, I never want to provide an excuse for racism – I'm terrified that someone might read this book and say, 'well, I've never had the chance to mix with Black people so it's OK for me to hold these views.' Ignorance and inexperience should never, ever be used as an excuse, or a justification for racism, homophobia, or any kind of hate speech. But can this cultural inexperience provide something towards an explanation, and the starting point for education?

I get myself in a complete pickle with all this. I want to share a voice from the world I grew up in; our truth. I turn to the one person who always helps me see the wood through the trees – my dad.

'It's about being small-minded,' he says. 'Looking inwards instead of outwards.'

Dad has lived in a small, all-white, farming community his entire life. He adores travelling but a combination of too much farm-work and not enough money has prevented him from seeing as much of the world as he would have liked. Instead, he watches hundreds of films and documentaries and enjoys studying maps and atlases. He taught me the capital cities of every European country when I was a little girl, and there's a spinning globe on his desk, next to the livestock movement records.

He's a classic 'boomer' – unapologetically traditional – and says controversial things mainly for the sport of winding us up. He'd probably call me and my sisters 'snowflakes' if he knew what a snowflake was. He's also a deep thinker; compassionate and empathetic and accepts people for exactly who they are. If Dad were a *Game of Thrones* character, he'd be a gruff man of honour – like Ned Stark. His opinion matters to me greatly.

When I share my ethical conundrum, Dad makes several points that give me a lot to think about. He reminds me that lots of rural people, especially farmers, lead very insular lives. He knows plenty who never leave their area, who have no desire to travel, and show zero interest in seeking out new experiences.

'They're good farmers, very knowledgeable,' he says, 'but they only ever want to talk about farming. It's all they think about. You try talking to them about anything else – they just switch off, they're not interested. They look inwards, not outwards. They only want to talk about what they know.'

He's also noticed a social awkwardness when conversation drifts from the familiar:

'I know plenty of farmers who wouldn't think of themselves as racist at all, but they're socially awkward and could quite easily say something that could be seen as racist. They wouldn't mean to offend – it would come from feeling uncomfortable.'

According to Dad, they're the ones who rarely leave the farm, or experience much outside the farming bubble.

'There's lots that do though mind,' he adds, not wanting to do his community a disservice. He uses his friend Dicko as an example, who's travelled the world shearing sheep.

'Now Dicko is interested in things,' he says, 'he's been to New Zealand and Norway and all over the shop and he's spent a lot of time with foreigners. He asks questions, he listens to people and learns. It's a different mindset somehow. That comes from broadening your horizons. Looking outwards.'

I mull this over. It makes a lot of sense to me, but I'm still left with questions. We both know there are people with racist views in our community – Dad doesn't deny it when I throw a few names around. Why though, I wonder? Where does it come from, having something against people you've never met? Resenting communities who live miles away?

Dad sighs and shakes his head:

'It's these little hierarchies. They get together in their groups, especially businessmen, and they want to impress each other. 'His business is doing well, and he's just said something about Black people, I want to work for him, so I'll agree with him,' that kind of thing. They don't start out racist, it happens without them even realising it. They get each other all fired up. They bond over it. Makes them feel part of the clique.'

'A lot of people are too scared to speak their own mind,' he adds thoughtfully. 'They just agree with whoever shouts the loudest, and they're usually the idiots.'

If I met someone like that in Bristol it would be very easy for me to freeze them out of my life and never see them again. But, as Matt Naylor pointed out, this simply isn't possible in small rural communities where everyone knows everyone, and you physically can't avoid people. They're in the livestock market every week, your kids are in school together, they prop up the bar in the village's only pub, you go to the same weddings, funerals, church services, or whatever. It breeds a tight-lipped stoic tolerance – putting up with offensive jerks basically. Very few people challenge them – avoiding confrontation for the greater good of smooth community cohesion. This kind of behaviour gives racism a pass – and it's the silent majority who need help, these 'passive bystanders' who see racism, and hate it, but do nothing to stop it. I've been that person; I know Dad has too. Somehow, we need to figure out a way of empowering the quietest, most conflict-averse person to call out prejudice within their own community.

Thankfully, Theeb the super vet is on it. He travels to village halls over the country, giving talks and running diversity training for small groups of people, in a safe, non-judgemental environment.

When it comes to embracing diversity, Theeb says there is a bell curve: 25% of people will never change, 50% can change or want to change and 25% are on the path to change.

There are three stages – first, you must identify the problem and acknowledge your own prejudice. Step two is being willing to change it. Step three is taking the action to change. Theeb rattles off a long list of organisations that are embracing diversity training: the Royal College of Veterinary Surgeons, British Veterinary Association, university vet schools in London and Nottingham, the Horse Racing Board, Pony Club, McDonald's, Oxford Real Farming Conference, NFU Education, Country Trust, National Trust, Museum of English Rural Life, Rural Diversity Network (Arts Council) and the Oxford Farming Conference.

'The conversation in rural and agricultural circles has definitely increased over the last couple of years,' he says, 'and real action is now needed as the next step.'

There are increasingly progressive voices speaking out on these issues – white allies too, putting their shoulder to the wheel to help make British agriculture a more multicultural and socially inclusive industry. Earlier, Theeb asked: 'who is the influencer for farmers?' Thanks to social media, we're seeing a new generation of farming influencers, people like Will Evans, creator of the Rock and Roll Farming Podcast. He interviews all sorts of inspirational people in the ag industry worldwide. He's your proper farmer and a proper progressive, and when Will calls something out, people listen. They respect his opinion and agree with him – more than that, they want to be seen agreeing with him. There was a time when young farmers, especially, looked to the old school macho, alpha-male role model – I'm not saying Will isn't macho or alpha – but there's more to him than a big tractor and a rugby shirt.

Groups like Black Girls Hike are inspiring people of colour to don their walking boots and explore the British countryside, to own it and feel at home there.

A few years ago, I started a small communications project called Just Farmers, aimed at building confidence among farmers and growers about sharing their stories with the media. In 2018, when I recruited the first group, there were only two women out of 12 farmers. By 2020, in the fourth group, there was an even split of men and women, and I didn't even have to try that hard to find them.

We still have a way to go. At time of writing, out of 60 farmers, there are only three people of colour and two from the LGBTQ+ community, but one thing I can say with absolute pride and certainty is I have never heard the slightest hint or utterance of racist or homophobic sentiment in a Just Farmers workshop. And we are proudly rural, uniting farms of all shapes and sizes from Cornwall to the Scottish Highlands, West Wales to Norfolk. I choose to take heart from that fact; that the community we've built is kind, caring and inclusive.

I do believe there is a genuine desire to get better. In 2022, the Oxford

Farming Conference rebranded its 'Emerging Leaders' initiative as the 'Inspire' programme, awarding 18 fully funded places at the (very pricey) annual conference to people who, as well as demonstrating their commitment to agriculture, are working to improve diversity in the farming world. One of them is Flavian Obiero.

I catch up with him over Zoom from his empty flat in East Sussex. He's in the process of moving home and changing jobs. He's leaving Plumpton College, after managing their 130-sow indoor pig unit for six years, and he's going back to the Hampshire farm where he first started his farming career. The same place where he got nicknamed 'darkie'.

'I'll be working with the same people, and I've already said to them, "Look, the stuff you used to say to me before, I won't stand for that now. Times have changed. I'm different." And one of them said, "Fair enough and we shouldn't have said those things anyway."'

'Are you satisfied with that?'

Flavian gives a relaxed shrug.

'I think people should accept that people make mistakes. There are things I used to say: jokes about women and LGBT communities because I was ignorant or trying to be cool. Thirty-year-old me now would slap the living shit out of 18-year-old me for saying stuff like that. But I have educated myself so I'm not like that anymore. I'm trying to get other people to do the same and change their ways.'

To help Flavian and Theeb shift up a gear, from conversation to action, we all need to join their movement. We all need to make a little pact with ourselves that we will call out racism next time we hear it. That doesn't mean pointing fingers, causing rows, or publicly humiliating our friends, family, colleagues, or neighbours. It means choosing the right moment, making sure we feel safe, and taking that person aside and having a quiet word, calmly and respectfully. Who knows, you could help someone get over the bell curve and on to the path towards positive change.

To the urban world:

You are lucky to have multiculturalism in your life – embrace it, explore it, and celebrate it. Because it is a gift – and not one that everyone shares. A rural child cannot help growing up in an all-white community. As they grow and learn, as I did, they may make mistakes and say the wrong things. Check your urban privilege and try not to judge others by your own lived experience. The countryside is changing and moving in the right direction, but simple demographics put it at a disadvantage compared to the cities. For that reason, some extra patience and understanding is needed in this space.

The rural working class are a disadvantaged minority too. Who speaks for them? Where is their movement? Where is their activism? In some ways I feel they are among the most voiceless of all. We say we sympathise; we say we're listening – but are we really? I've tried to show their truth in my journalism – but so much of that truth gets edited out (and I'm not referring to *Countryfile* here). Ugly pictures erased because it looks like 'poverty porn': a farmer who smokes, edited out because audiences don't sympathise with smokers; a roll of rusty barbed wire in a working farmyard edited out because it looks 'messy'. We sanitise the rural working class, romanticise their struggle. But when the ugly stuff breaks through, like shit bursting from a broken pipe, we turn away in disgust. We're revolted. The moment they say things we disagree with, we shut our ears off – as if their views are catching, terrified of being tainted by their prejudice. I don't want to turn away anymore. I want to listen, and respond: 'Why do you say these things? Where is it coming from?'

To the rural world:

Perhaps you think I'm just another middle-class urban snowflake pushing cancel culture on to country people who make a few jokes but mean no harm? Maybe you think diversity isn't really a rural issue? After all, it's not our fault the countryside is mainly white, and what's wrong with that anyway?

I've heard it all before.

The reason rural Britain needs to change is because it is missing out. If it stays the way it is, it will get left even further behind while the world around us changes and moves forward. If we want rural communities to be thriving and relevant for the next generation, if we want to attract the best minds, the best investment, the best entrepreneurs we need to change. Diversity isn't about politics, or racism, or horrible things – it's about people and progress and positivity. But some outdated attitudes have found refuge in rural areas, and they are standing in our way. Try Googling 'rural UK racism' – it's depressing. I couldn't look beyond page four. That's the image rural Britain is projecting to the world. It's ugly and we need to change it – and fast.

You may not have the gift of multiculturalism in your life. Please do your best to find it, somehow. The rewards are huge. We know diversifying a farm makes good business sense. We know diversifying crops makes good environmental sense. Well, diversifying our communities makes good cultural sense. Diversity is never a bad thing, in any context.

Jane Hayes-Hodge at Rise & Root Farm in New York gave me a wonderful piece of advice: 'Get comfortable being uncomfortable,' she says. 'We as white people have been trained to not make people uncomfortable and I think we need to start liberating ourselves from that. We need to build relationships with people of colour. If I live in a rural area and all the people I talk to are white – I need to go somewhere and meet people that I'm not used to and don't feel comfortable around. I need to find them.'

And it doesn't necessarily have to be that literal. I asked Conrad about it, our Jamaican allotment neighbour:

'You don't have to go out of your way to meet people,' he says. 'I was born in Jamaica in 1969, I never knew that I was going to meet you, sit down at a table with you on an allotment and have this conversation, but it's part of my destiny, and I'm part of your destiny – that's why we are here today. When you meet me, and I meet you, what do we do in that moment?'

'Say hello?'

'That's right. We could say, 'oh, he's Black' or 'she's white' and ignore each other and just get on with our business, or we could say 'hello' and have a conversation and get to know each other.'

I smile in the spring sunshine. It all starts at 'hello'.

CHAPTER FIVE

ANIMALS

IN THE DAYS LEADING UP TO the winter lockdown, in November 2020, I snuck home to the farm for a few days, squeezing in one last visit before we were separated again for many more weeks. I spent a day helping Dad. We did the usual morning rounds and drove down the road to check on the sheep in the quarry, 38 acres which Dad rents off his younger brother, my uncle Stephen.

It's an old limestone quarry which shut down in the 1970s. Grandad Bill used to work there on the stone crusher, making sure rocks didn't get jammed. Dad remembers George the quarryman passing the farm on his way to and from work, hobnailed boots clattering down the road at dawn and dusk. You can still see the old lime kilns, now overgrown with ivy, and a rusty steel and tin shed on concrete plinths where crushed, powdery lime poured down the metal chute into the wagons waiting below. Once loaded, they'd chug on to the weighbridge and check in at the pokey office, no bigger than a garden shed, where Nan Beryl's younger sister, Anne, would check and stamp their tickets. The office is still there, though slowly being consumed by scrub. I step inside over prickly brambles, dead leaves crunching underfoot. I peer over the wooden counter, picturing my auntie Anne sitting at a little desk on a dark winter's night, an electric two-bar fire warming her feet.

The quarry is a ghostly place, abandoned and frozen in time. I struggle to picture our sleepy little hamlet as a place of busy industry – horns blaring, dynamite blasting, rocks thundering into the crusher and a steady procession of heavy trucks coming and going all day long. It's like they just shut up shop one day, turned off the lights and left. Decades of manual graft now almost totally rewilded and reclaimed by nature.

I can tell Dad hasn't thought about the quarry in a long while. He's too busy farming to dwell on what was here before, but I catch his eyes grow soft, and distant, as remembering takes hold. There are many who would gladly consign our industry to the history books too. A father and daughter could walk this land in the twenty-second century and say, 'They used to farm livestock on these fields – cows and sheep would have roamed here once.'

Could that really be?

We trundle down the sloping field. Gyrn Moelfre and the Berwyn Mountains beyond are looking mighty fine today, laid out before us in full panoramic gorgeousness. I adore these rolling hills. I drift off in the passenger seat of Dad's Toyota Hilux, doing what I always do – trying to absorb the beauty and imprint it on my mind. It's become a bit of an obsession this, trying to capture the view and keep it. Get in my brain. Get in there and stay there! I'll need you in Bristol when I'm looking at fly-tipped mattresses and dog poo. While I'm mid-gawp, Dad slams on the brakes, jolting me into the moment. A lifetime of patrolling fields of livestock has honed a hawk-like observance for anything unusual or just slightly off. He's spotted a sad and sorry looking sheep pressed up against the fence along the hedgerow. I would have driven straight past, lost in a daydream.

It's one of his new tups. A young Beltex-cross-Texel ram. It's the height of breeding season so this young buck should be running with the ewes having the time of his life. Instead, he's all alone, stooped and shivering with his head down in the cold and drizzle. He's smaller than the tups Dad usually buys and looks a bit on the thin side. 'Oh dear,' says Dad, 'he's gone down-hill. I knew I shouldn't have bought him when I saw him in the ring.'

He reckons he'd been 'dumped' on him at a ram sale by an over-zealous auctioneer but, as a new buyer, Dad felt obliged to take the little chap home, paying £220 for the trouble.

As we park up and wander over for a closer look, the miserable-looking tup doesn't even try to run away. He just turns his head, barely a whisker, for a better look at us. 'We'll have to take him home,' says Dad.

We lift him into the back of the truck. I've got his back end while Dad hoiks the front half of his body on to the tailboard. He's surprisingly heavy. 'Careful he doesn't kick you in the face Anna.'

I'm doubtful. He can barely stand up, let alone muster the strength for a roundhouse to the face. He slumps down in the corner.

Dad climbs back into the driver's seat: 'We'll put him in the shed, get him warm, give him some feed and he should perk up in a couple of days.'

We check on the rest of the sheep and, on our way back up the field, I turn round to see how our invalid is doing. He's gone. The back of the truck is empty. In some miraculous show of strength, like a wounded action hero, the little tup has thrown himself out of a moving vehicle. 'Stupid bugger,' sighs Dad.

We turn around and retrace our path and find the tup in a crumpled heap on the ground. He looks up forlornly, as if to say: 'I don't know why I did that.' We pick him up again and roll him back into the truck. This time Dad ties three of his legs together with baler twine. He looks an even sorrier sight.

Back home, we untie his legs, lift him down from the pick-up and gently guide the wobbly little fella into a pen of fresh straw. He totters over to a bucket of water and gives it a half-hearted sniff. Sun is streaming through slats in the side of the barn, casting golden stripes across the floor. The cows look on curiously. Dad quietly observes the patient and frowns when he notices his rapid, shallow breathing – a sure sign of pneumonia. He gives him a shot of antibiotics and holds a scoop of feed to his mouth, waiting patiently as the little tup nibbles away. At least he's eating.

That's all we can do. It's not worth calling a vet. It's down to him now. We head off to put up some electric fencing.

The following morning I'm sitting at the kitchen table, eating a bowl of cereal, chatting to Mum. Dad walks in the back door and kicks off his wellies. 'That tup's dead,' he announces unceremoniously.

'Oh no, poor thing.'

It's sad. I feel genuinely sorry for the little tup on the wobbly legs

– but I don't dwell on it for long. I move on with the stoicism of the desensitised.

Later that morning, driving back to Bristol, I get caught in traffic on the M32 entering the city. There's a large sign on one of the footbridges. It says: 'Animal Farming: House of Horrors'.

There's been a Go Vegan sign on that bridge for several years now. It is yet another stark reminder of the difference between my urban and rural worlds. I picture that sign hanging from a bridge in Oswestry, in the heart of a farming community. It would cause absolute uproar. It would probably make the front page of the *Oswestry and Border Counties Advertizer*. But here in Bristol, it's part of the furniture. You barely notice it. You can't swing a cat (gently) without hitting a vegan. I read the sign with the stoicism of the desensitised.

I ponder this latest anti-meat proclamation looming overhead as I wait for the Bristol traffic to inch forward. Is our farm a house of horrors? I've seen lots of dead and dying animals over the years. What happened to the little tup was pretty horrific – catching pneumonia in a cold, wet field, getting chucked in the back of a truck and hurling himself out again (no doubt terrified) and dying alone at night in his pen. Does that make Dad a horrible person? Am I the horror for standing by and being OK with it?

And then I remembered something else Dad said that morning, after he'd discovered his tup had died: 'They break their hearts easier than the ewes.'

He meant male sheep give up easier, that they lack the fight of the females. I've often heard my parents say that – I'm not sure if it's a scientific fact – but the sensitivity in his words struck me: 'They break their hearts.'

He had spotted a lone, heartbroken sheep and brought him in from the cold. Would my dozy untrained eye have picked out the speck of white cowering in the hedgerow? I doubt it and how awful that would have been – to leave him there, trembling, waiting for death. I picture Dad busying around in the shed to make the little tup comfortable – putting down fresh straw, giving him medicine, gently holding a scoop of food to his mouth. That image of Dad, silhouetted by rays of sunshine pouring into the barn, bending down to help one weak sheep eat: far from horror, I saw tenderness.

Is 'animal farming' a horror? Or is life and death the real horror? That mortal inevitability which plagues us all – just as humans live and die, so too do animals. Perhaps.

But when it comes to farmed livestock, there is another inevitability. A discomforting crux which is impossible to avoid: had the little tup survived, he would have died eventually – and at our will. Dad wouldn't have kept him for another breeding season; he didn't make the grade, so he'd be off to the abattoir and slaughtered. This fundamental fact is where the fundamentalists get farmers on the ropes. Because it's true. We'll kill them all eventually. And eat them.

But is it wrong?

No one can tell you if it is right or wrong to eat animals. It is a decision we must come to by ourselves, as individuals. I believe we should all spend time pondering that question, consciously and with respect, and check in with it throughout our lives. Am I still OK with this?

One evening in lockdown, as Alex and I eat dinner on our laps in front of Netflix, I ask him:

'Why do you think it's OK to eat animals?'

He doesn't have to think about it for long:

'I believe it's my right as a human being. I'm at the top of the food chain.'

I'm surprised by his directness – I hadn't expected such a straight answer – and I slightly envy his surety. There isn't a hint of doubt.

I've had doubts. There are moments when I look at our cows and sheep, oblivious to their fate, that I feel guilt, even a stab of shame. There's the uncomfortable paradox of bringing baby lambs into the world knowing their sole purpose in life is to die. We produce commercial lambs for the meat market; they are not breeding stock, and most will be slaughtered before their first birthday. There are a few exceptions – Dad kept my sister's pet ewe lamb for breeding, on strict orders from Mum that 'Milly mustn't go to market'. Milly remained part of the flock until she was an old lady and never forgot my sister Kate. She'd even wander over to the fence every lambing season to show off her new babies.

The philosophical debate whirring around animal agriculture and meat-eating is so emotive and wide-ranging, pulling me down umpteen rabbit holes, that the simplicity of vegetarianism or veganism has been a tempting option at times. Cut out the meat to cut out the noise.

I listened with great respect to Peter Singer, author of the 1975 book *Animal Liberation*, on the PETA Podcast. He pioneered the philosophy of speciesism – that our exploitation of non-human animals is based upon a misguided belief that one species is more important than another. Humans arrogantly place themselves at the top and then select which creatures get to be their friends, like dogs, and which will be their food, like chickens. The simple act of eating other animals, according to Singer, rests at the very toxic heart of speciesism.

There has been an animal rights renaissance in recent years, and it's entered mainstream consciousness more than the protests I remember on the news in the 1990s, grainy recollections of people in dark jumpers waving placards against animal experimentation and live animal exports. Maybe I'm not quite old enough to remember the societal impact it had, but I'm not sure they spawned as many plant-based cafés or blockaded McDonald's depots in bamboo treehouses. The modern vegan movement is so much more than a dietary choice – it's a wide-ranging code of ethics encompassing animal welfare, environmental concerns, and the fight against climate change. It has captured the spirit of our age perfectly.

Personally, I'm grateful for their activism and, contrary to popular belief, I've heard many farmers say the same. It means people are finally engaging with their food system, which is much better for farmers than the status quo of being ignored, taken for granted, and shoved in a dark corner of public consciousness. It has pushed the animal welfare agenda forward too.

In May 2021, the UK Government introduced the Animal Welfare (Sentience) Bill, which formally recognises, for the first time, animals as sentient beings who can experience joy, pain, and fear. The debate has applied pressure on the livestock industry to raise its game and it's made consumers, like me, think much harder about the food we eat. Sure, some vegans are smug, shouty, self-righteous and really, really annoying: 'How do you know someone's a vegan? They tell you.' But, ultimately, it's a good thing. Because we needed to hear it.

I've listened to all the debates but, in the end, my reasons for sticking with an omnivorous diet are health-based and anthropological. I like the sense of evolutionary connection to my hunter/gatherer ancestors and the generations of livestock farmers in my genes. I have great respect for the symbiotic relationship between humans and animals and find the evolution of domestication over thousands of years utterly fascinating – how we tamed the wolf and bred the dog to be our loyal companion and broke wild horses so we could ride them. Cats, cows, sheep, pigs, poultry – all sorts of creatures have grown dependent on us for their protection and survival. And we look after them in return for many things – companionship, work, sport, meat, milk, eggs, leather, wool. In my mind, it's a partnership we can't just walk away from. We sustain each other. Be it a pet or a farm animal – we have evolved together, and we are inextricably linked. For me, there is reverence in farming animals and eating meat.

And I do believe the UK has some of the highest animal welfare standards in the world – not necessarily 'The Best World Leaders in Welfare Ever . . . EVER!' or whatever self-congratulatory superlatives we seem to love chucking around. We're far from perfect, but we are pretty darn good. The first ever law protecting animals was passed 200 years ago right here in the UK: the 1822 Act to Prevent the Cruel and Improper Treatment of Cattle. Our very own Royal Society for the Prevention of Cruelty to Animals (RSPCA) is the oldest animal welfare charity, founded in 1824. I've visited enough farms around the world to say with confidence that were I a farm animal – I'd choose to be a British one. I've seen calves branded, dehorned and castrated without anaesthetic in the Midwest of America and the Australian outback; I've seen animals limping and fly-bitten on smallholdings in Africa; I've seen female pigs biting the bars in American sow stalls (that's before they even get to the farrowing crates); I've seen housed dairy cows drinking urine off the floor in the Czech Republic (indicative of a salt or mineral deficiency) and on a Danish pig farm I felt my way along the pens with weeping eyes clenched shut – not out of shock or disgust, but because the ammonia stung them so badly.

On the health side, for me personally, I need animal protein. Not in huge amounts and I certainly don't eat meat every day, but I have noticed my body doesn't respond brilliantly to a completely vegetarian diet. Whether it's iron, B12 or simply psychological – it works for me.

So, if we personally decide it's OK to slaughter and eat animals, from here on in, the welfare debate gets much more complicated, because looking after an animal's best interests in life, and at the end of life, arguably poses more ethical questions than simply deciding whether or not to eat it once it's dead. How can we really know what they want from their lives? We are humans, not farm animals.

The 'five Freedoms' act as our guiding light. These are internationally accepted standards, first developed, again, in the UK in 1965 and have evolved slightly to encompass the mental as well as physical needs of animals. They are:

1)   Freedom from hunger and thirst.
2)   Freedom from discomfort.
3)   Freedom from pain, injury, or disease.
4)   Freedom to express normal behaviour.
5)   Freedom from fear and distress.

That all looks clear enough, until you realise the freedoms are open to interpretation and can even contradict each other – especially when you compare indoor and outdoor livestock systems. For instance, an indoor cow can't graze on grass – restricting its normal behaviour – but an outdoor cow can pick up worms and parasites, causing disease. An indoor pig can't root in the soil – again, normal behaviour – but an outdoor pig can suffer heat stress in summer, leading to discomfort. It's a minefield. My journalist trump card, which I've whacked out a few times during interviews over the years, would say: 'Ah yes, but the outdoor animal is more natural, they've evolved to be outside, so surely that's better welfare?'

I try this out on Theeb, who makes another fundamental point:

'With farmed animals we've created breeding objectives, so we've taken away a lot of their natural characteristics. A Holstein dairy cow would be dead within a couple of days if it were in the middle of the forest. They're not "natural" animals, they're farmed animals.'

We're back to that symbiotic relationship. We haven't just domesticated wild animals, we've created breeds within species, and crossbreeds within sub-species, and all with their own unique set of welfare needs. What works for one cow might not suit another. Take Holsteins – they've been bred to produce eye-popping quantities of milk. Theeb says: 'The energy needed to produce the 10,000 to 14,000 litres a year they do needs the same calories a day as we would need if we were to run a marathon every day. That's why many Holsteins are kept indoors: grass on its own will not provide enough energy and protein if we are to optimise milk production from these animals.'

A Jersey cow is a different kettle of fish. They're smaller and produce less milk, meaning they do much better on grass. They can handle being left in a field with nothing else to eat and still produce good quality milk.

So, a happy Holstein does not a jolly Jersey make.

'Unfortunately, when we farm an animal, we take away some freedoms,' says Theeb. 'Even if you put some deer outside behind a fence, or in a sanctuary, you've restricted their freedom. Good welfare is finding the best way of supporting the animal in a farmed system.'

We're going deeper down the rabbit hole of welfare now. Assuming we've decided it's OK to kill and eat animals, and we've agreed to restrict their freedoms in order to farm them – what next? Now we're faced with a myriad of different farming systems to get the same end product – our milk, eggs, cheese, pork, chicken, beef, lamb. We can choose from a big pile of labels: free-range, organic, grass-fed, outdoor-bred, outdoor-reared, Pasture for Life, Pasture Promise, Leaf Marque, RSPCA Assured, and Red Tractor. The latter is the industry benchmark. It has come under fire for being too much of a catch-all, and even letting a few sub-standard farms slip through the net. Personally, as a shopper, I use it solely as a guarantee of Britishness. That's important to me for all the reasons I stated earlier, especially as we face the prospect of more imported meat, potentially reared to lower standards than our own, popping up on supermarkets shelves because of all these new free trade deals the Government is drooling over, post Brexit.

So, labels tell us something about how, or where, the animal lived. As a rule of thumb, the more 'natural' its life, the more it's going to cost you – and that higher price often has more to do with growth than welfare. Outdoor animals take longer to 'finish', or fatten, than indoor animals because they use more energy walking and running around and their food, usually grass, hay, or silage, isn't as calorie-packed as the grain and protein blends plonked in front of animals' noses in a pen or barn. Effectively, you're paying for the ultimate welfare win – a longer life. But can you tell from a label if the animal was 'happier'?

'The problem with welfare is it's being used as a marketing tool,' says Theeb the farm vet. 'There are all these different assurance groups using welfare to compete against each other: "We're better than them because we measure welfare in this way" and so on, when, really, they've just got different standards. But to the actual animal, does it make any difference? I don't know. Sometimes it's more about competing with each other than helping the animal.'

I visit lots of farms, of all different shapes and sizes, and spend considerable amounts of my own time pondering indoor versus outdoor production systems (genuinely, these are the sorts of things I think about while washing the dishes or walking the dog – I need to get out more). I often wonder to myself what the human equivalent might be? In my mind it's a choice between being locked in a hotel room where someone brings you food several times a day. It's temperature-controlled, cleaned daily, and someone caters to your every whim (providing you're in a good one – if you're unfortunate enough to end up in a bad hotel, with horrible staff and terrible facilities, you're stuck with it). You get to share a room with your mates but there's nothing to do – no gym, no smartphones, TV, or Netflix – and you can't leave. You just eat, sleep, and chill. For the rest of your life. Or you can choose a never-ending camping holiday where you're free to explore the great outdoors with your mates (providing you stick to the designated area) and you can relax in the sunshine every day. But you find your own food and if it gets cold, or rains, you can't come indoors. You stay out there, whatever the weather. For the rest of your life.

Humans would struggle in either scenario – it's a ridiculous analogy – but farm animals have been bred to suit these environments; high-maintenance pigs for indoor systems and low-maintenance pigs for outdoor systems.

At this point in our journey down the welfare rabbit hole, assuming the farms are well-run, and the animals are treated with care and respect, I'm personally OK with the fact some live indoors and some live outdoors. Some even get a bit of both. But here, our journey gets complicated. We've entered a chamber with countless passages heading off in all directions: Pigs this way! Please choose from farrowing crates or freedom-farrowing pens. Tails docked or bitten? Dairy this way! Calves in hutches, calves in pens or calves at foot? Poultry this way! Please choose from caged, barn, or free-range? Beaks trimmed, or feathers pecked?

It is dizzying. Animal welfare in modern farming is a confusing rabbit warren of conflict and contradiction. There seems to be a trade-off at almost every turn.

A piglet's tail is cut to stop it getting bitten by other pigs, the sharp bit is shaved off a chick's beak to stop it pecking other chickens – the idea being a brief moment of pain when they're young prevents the risk of much worse pain further down the road – but mutilating animals creates a whole other welfare issue and doesn't address the fundamental reasons why they're cannibalising each other in the first place: usually boredom and confinement.

Many newborn dairy calves are put in hutches or individual pens to protect them from disease, but herd animals don't like being alone. Other farmers prioritise socialisation over disease risk and prefer to keep calves in small groups. Whichever way you choose to do it, calves are still taken off their mothers in their first few hours or days of life. Calf-at-foot dairy production, which keeps the cow with her calf, is growing in popularity but is unfeasible for most commercial herds.

Sows who have just given birth are kept in farrowing crates to protect their tiny piglets from getting squashed; a very real risk when mum is 200 times larger than her babies. The problem is, she can't turn around or move freely. There is mounting pressure to get farrowing crates phased out in the UK, just as sow stalls were banned in 1999, but the industry is pushing back hard.

Piglet mortality isn't just a welfare problem, it's bad for business. Fewer pigs means less profit. There are alternatives for indoor producers, such as freedom farrowing pens, where the sow is never constrained, and '360' pens, which I've seen used on a farm in Dorset. To an untrained eye, they look almost identical to a conventional crate but can be opened and closed, like 'accordion bellows' says the farmer, Robert Lasseter. He saw it as a good halfway house for his 250 sows and invested £200,000 to convert in 2014. At the time '360' pens were approved under the RSPCA Assured scheme as a higher welfare option, which earned a price premium. However, the RSPCA has since increased the pen size required, which means Robert's system no longer conforms to their rules – a considerable loss of income. He was also in the Co-op's Pork Producer Group until the retailer changed their rules stipulating all their pork must be outdoor bred. It's the perfect illustration of how our understanding of animal welfare is constantly shifting and evolving, and how farmers are often left playing catch-up. For many it feels like two steps forward, one costly step back. You solve one problem and another pops up somewhere else. It's welfare whack-a-mole. Domestication led to intensification, and intensification led to cheaper food, and cheaper food led to higher demand, and higher demand led to more competition, and more competition meant tighter margins, and tighter margins meant saving costs, and saving costs led to even more intensification and before we knew it, our brains were fried and we all split off down different welfare rabbit holes. Now we're shouting to each other in the dark: 'Wow! The animals are really happy here!' 'Rubbish! They're much happier over here!'

I don't know about you, but I've been lost down the animal welfare rabbit hole for a long time. I don't have the uncompromising dietary beliefs of a vegan but lack faith in the wisdom of all these 'efficient' livestock production systems. I'm fundamentally uncomfortable with a lot of stuff, and I believe the urban/rural divide has made it worse.

There are uglier aspects of farm life that I simply do not discuss with my friends in the city because I know how it looks, out of context. The simple fact, for instance, that there's usually a dead animal somewhere on the farm. Commercially bred sheep have a frustrating tendency to die in the most benign situations – from getting stuck on their backs to eating grass that is slightly too rich for them. Survival is not their strong point. It has also been illegal to bury dead stock on farms since 2004. Carcasses are now collected under the National Fallen Stock Scheme, by the knacker man who drives around the countryside in a stinky lorry. It can cost up to £30 per animal, so cash-strapped farmers usually wait to build up a bit of a pile to get a multiple pick-up discount.

I prefer to avoid the subject of dead sheep but sometimes it almost smacks you in the face. On my 26th birthday, my friends from Birmingham came to the farm for a barbecue and camping. It was a lovely weekend, drinking beers on the lawn in the sunshine, Mum and Dad buttering baps on the garden table as burgers sizzled away. Excited by all the company our border collie, Cassie, decided to bring our guests a welcome present. She trotted on to the lawn, tail wagging happily, and proudly dropped a dead lamb's leg at the feet of my Londoner friend Steve. 'What's that?' he said, bending down for a closer look, beer in hand. 'Eurgh! It's got a hoof!'

I was absolutely mortified. I wanted the ground to open up and swallow me. Steve found it quite funny. He wasn't offended in the least.

'Of course I wasn't,' he said years later, 'I don't expect the countryside to be some kind of sanitised environment.'

My reaction had more to with my own insecurities and assumptions, that urban people are somehow more squeamish than us country folk: 'You can't handle the truth city boy!' I turned out to be completely wrong. The divide can come as much from self-consciousness on the inside, as judgement from the outside.

Having said that, I fully appreciate that, to some people, my family and I are speciesists who exploit non-human animals for our own financial gain, and would no doubt call me a complete hypocrite for giving our spaniel Lucy a spoilt life of comfort in comparison to the functional existence of our farm animals.

One of the greatest ironies I encountered as a dog owner was on Robert's farm when I reported on tail docking for BBC Radio 4. I watched Robert cauterising piglets' tails with a hot curved blade. It wasn't as grisly as I expected but it was uncomfortable, nonetheless. Afterwards, as I walked back to the car, his spaniel puppy bounced over to greet me – with a long swooshy tail.

'You dock the pigs, but not the dog?' I observed.

'Yes, dog tail length is much more about fashion,' he replied, 'but we wouldn't stop tail docking pigs because of the welfare implications of not doing so.'

And here's the wallop of irony – our pampered city pooch had her tail docked as a puppy. It's not something we would have chosen for her but, by the time we met the litter, it was already done. She has no fear, trauma, or pain, happily lets us touch her tail and we've been told it prevents it getting caught and bloodied in prickly undergrowth. Crucially, she still has plenty of tail to wag and that's the most important thing. Suddenly my head was all over the shop – how can I judge this practice in pigs when my own dog has had it done?

So, I do hear the speciesist argument. I give it lots of thought and I do wonder how history will judge us. Maybe I will go vegan one day. Maybe I won't. All I have right now is my truth, which has been shaped by my lived experience. I believe you can love and care for animals even if you farm them and eat them. That is my truth.

But can you kill animals for fun, for sport, and still call yourself an animal lover?

I grew up hearing the distant crack of gunshots from the pheasant shoot across the valley. There's a game farm just down the road from Mum and Dad's and lots of grumpy signs in the local woods next to the pheasant rearing pens: KEEP OUT! DANGER! NO TRESPASSING! Hunting and shooting have always felt like closed off worlds to me, despite my rural upbringing. Grandad Bill was a beater, but I didn't get to see it for myself, and Dad used to shoot a bit, but never caught the bug. His gun sits idly in the corner, reserved for the odd rabbit. The hunt used to ride across our land years ago but, according to Dad, they never did much good:

'They only catch the old, crippled foxes,' he'd say. 'It's the young fit ones that do all the damage to the sheep – you're better off getting the shooters in.'

Dad used to go out with a huntsman's daughter when he was at agricultural college in the 1960s. He liked her family well enough but didn't care much for the hunters, who he described as a 'stuck up lot' who 'looked down their noses at everyone else'.

But I've always been curious about their world.

It's May Day 2021 and I'm drinking coffee at a kitchen island with a young female hunter. She lives in her late grandparents' bungalow, just down the road from the farm where she works. She's made it younger and trendier, adding girly touches here and there. There's a teacup and saucer, obviously a gift, saying Gin Queen in whimsical italics. On the windowsill behind the sink, as if to back up the statement, pink gin bottles lie in wait of a party. A vase of white lilies scents the air. Next to smiling photos of the girl, who I have decided not to name, are three matching mugs, each with an inscription: I'd Rather Be Drinking Pimms. I'd Rather Be Eventing. I'd Rather Be Hunting.

She has loved horses for as long as she can remember. At the age of four, on a family holiday to the New Forest, she spotted a wild pony in the distance. 'I want to sit on that,' she said to her Dad. They got close enough for him to lift his tiny daughter and place her on its back.

'Four years old and you sat on a wild horse?'

I'm incredulous. She giggles.

'Yes. No fear.'

By the age of 11 she was competing in dressage, cross-country, and show jumping – the three disciplines which comprise eventing. At 13, she had joined the Hunt, riding her pony Smudge up front with the huntsman, jumping on and off to open gates and gradually learning the names of 80 seemingly identical hounds:

'Every time I went up front, I had to learn the names of two new hounds and then pick them out the following weekend. I never managed to learn them all. Just like farming, being a huntsman is a way of life. The hounds are their flock basically and they know every single one.'

The huntsman was an 'old boy' who retired when the Hunting Act came into force in 2005, which made it illegal to hunt wild mammals with dogs in England and Wales.

'He just didn't see how he could carry on doing what he loved, in the way he knew how.'

She, on the other hand, was more willing to adapt and went trail hunting instead, where an artificial scent is laid down for the hounds to follow.

'You've got the low-key hunts,' she explains. 'They're what I call your "happy hackers" who love their horse and just want to have a nice jolly day and meet their friends. You don't necessarily jump. Whereas there are other hunts with people who are London-based. They have their horses on livery, come up for the weekend and have their horses delivered to them. They want the adrenaline and the hedge jumping, the big days and the accolade. It's a completely different ball game – you've got your money packs and the happy hacker packs.'

I wonder if it's changed much since the ban?

'From a jumping point of view, it's the same thrill,' she says. 'Drag hunting has always been there for the thrusters who are at the front jumping anything. The nutters basically.'

'Hounds?' I assume.

'No, riders! They'll jump eight-foot hedges with ditches on either side and hanging five bar gates and ridiculous things. Those are the people who could be spending £10,000 to £15,000 on a horse and have another five waiting in the stables for when they break one. It's basically having too much money. It's impressive until something goes wrong. I'm somewhere in the middle. It is such an amazing feeling having that partnership with a horse, when you can point them at a hedge and they just go for it, but I don't like jumping gates. I'll open it thank you very much. These things open if you didn't know!'

We laugh. I enjoy chatting with this girl. I find her incredibly down-to-earth and easy to talk to. It's probably why I'm full of questions.

'Are you satisfied with trail hunting then?'

'I am ...'

She pauses. There's a 'but' coming.

'But watching a pack of hounds hunt a fox and then catch it, there is a thrill there. It's a little bit like ...

She pauses again. She already knows I won't understand, and searches for the right words. She levels with me:

'I see life and death every single day. When I was a child, and hunting was still legal, I was blooded.'

'What's that?'

'So, when a fox is killed ... this is barbaric ...'

She looks away and laughs nervously, shifting in her seat.

'As a child you get the blood of the fox put on your head.'

'What? Like some kind of initiation?'

'Yeah basically. And seeing that from such a young age, it does just become normal really, which a lot of people will never understand.'

I can't hide my shock and she sees it on my face. I can't say I understand, but I am curious. I want to learn more. I want her to continue sharing with me.

'Because I grew up with it, it isn't demonised,' she says. 'It's not that I don't feel bad about it, I know we're responsible for destroying a life, but I can see the bigger picture. I've seen plenty of foxes that have been shot and wounded and, to me, that's far more barbaric than a pack of hounds ripping a fox apart. It's instant.'

'Does it still happen? Are the hunts still killing foxes?'

'Yes.'

'And how does that make you feel?'

'I kind of, just, ignore it.'

'You turn a blind eye?'

'Yes.'

'Is it justifiable to kill a fox with a pack of hounds?'

'I think it can be. Yes.'

'Is it justifiable to break the law?'

A long pause.

'No. It's not justifiable to break the law.'

Her view, in a nutshell, is simple: she would like to see the ban lifted and believes hunting with dogs is a good form of pest control, but she's accepted that's never going to happen and wishes the rest of the hunting community would wrap their heads around that too.

'It's been a blatant disregard for the law for the last 17 years,' she says.

'And that troubles you?'

'It gets to the point where we become a laughing stock. The amount that does actually go on . . . if they really wanted to get hunting 100% stopped and banned, there would be plenty of evidence to go and dig up and get the job done.'

She goes on to describe a culture of deep denial within the hunting world. She finds it baffling how, in her view, hunters and huntsmen have cast themselves as the victims – truly believing they are the ones being wronged:

'You can't hold your hands up in horror when the police are stopping what you want to do, because it is illegal!' she says in exasperation. 'At the end of the day they are breaking the law!'

She agrees when I suggest institutionalised arrogance – not the ban – will bring down their sport.

'The future of hunting isn't looking great,' she shrugs with a look of benign acceptance. 'It's 17 years since the ban and the sabs [saboteurs] are getting more violent. It's getting dangerous. We might not have any of it in the next five years, which I think is a very realistic thing to start thinking. It's dying out.'

She is incredibly upfront. For years, the hunting organisations have insisted, officially, that they are obeying the law and mistakes are accidents, and accidents are mistakes. It backfired in August 2020 when recordings of Mark Hankinson, a director of the Master of Foxhounds Association, speaking to fellow hunters in private webinars, were leaked online. He was found guilty in October 2021 of encouraging illegal foxhunting and advising huntsmen on how to evade the ban by using trail hunting as a 'smokescreen'.

The girl holds her head in her hands:

'Jesus Christ you idiot, what have you done?'

I need to understand what separates her from the diehard deniers within her community. Where does this upfront honesty come from? She thinks about it, looking out of the window with her hands wrapped around her mug of tea.

'I'm a lot more open-minded than some of my friends,' she says. 'I've travelled and I've seen the world and a lot of your hunting people haven't even left the country. They are very closed-minded.'

She may hold different views in private, but she, like so many of us, will bow to peer pressure in public. She will look the other way when the hounds catch a fox. She belongs to a culture. It's her home and everything she has ever known. I ask about her own complicity:

'I will speak my mind but there comes a point where they just won't listen or don't want to listen and that is one of my social circles – I don't want to be alienated from it. There are a lot of people my age who are still diehard hunters and, if I was talking to them right now, I would probably come across as a diehard too.'

I suggest to her that hunting foxes with dogs is no different from cock fighting, badger baiting or dog fighting – except those are traditionally the blood sports of the working classes. She doesn't deny the ingrained snobbery and says many hunters would see themselves as being 'above all that'. History, she suggests, may have given foxhunting a pass – the association with kings, queens and aristocracy has given one blood sport a refined invisibility cloak over another.

She disagrees with only one challenge I throw at her – the class thing. I find it fascinating how she accepts illegality, arrogance, and even snobbery – but draws the line at poshness.

'That perception is definitely wrong,' she says.

She is not a toff – she went to a state school and has a broad country accent.

'There are a few toffs who sound a bit plummy,' she says, 'but I'd say it's 20% toff and 80% normal people. Huntsmen often don't have two pennies to rub together – they do it for the love of it.'

'A bit like gamekeepers?'

'Exactly.'

I feel deeply thankful for her honesty. It has given me an unfiltered insight into a world I knew nothing about. I still can't say I understand it – or agree with it – but, the surprising thing is, I believe her when she tells me she's an animal lover.

Very few people feel nothing for animals. We truly are a nation of animal lovers, but the welfare debate is an emotional spectrum of perception. One person's cruelty is another person's prevention of cruelty. What we perceive as good or bad animal welfare often comes down to what we as individuals are comfortable with, and that will almost always be influenced by our own personal level of sensitisation, or desensitisation, to the issues. Farmers and those from farming families become very familiar with livestock and the cycle of life and death – we see a lot of stuff – which can contribute towards varying levels of desensitisation. 'If you've got livestock, you've got deadstock,' says Dad matter-of-factly.

People from non-farming or urban backgrounds, who haven't spent much time on farms or talking to people in the farming community, may glean information from books, newspapers, film and TV, social media, or campaign groups. This can contribute to varying levels of sensitisation. Take London-based journalist Olivia Rafferty: she went pescatarian at 12 years old after watching the 2008 documentary *Food*

Inc. which looked at the inner workings of America's corporate food industry:

'I was absolutely traumatised. I never wanted to eat another living thing ever again.'

She turned vegan when she was 18, straight after watching *Cowspiracy* – the notorious eco-exposé which explores the impact of animal agriculture, mainly cows, on the environment.

Farmers and non-farmers are learning about animal welfare in different ways; one isn't necessarily better than the other, but they come with inherent biases and blind spots. The problems arise when people believe they are 100% right and plonk themselves at opposite ends of the spectrum of perception – with furious farmers at one end and angry activists at the other – making it more difficult to achieve anything constructive which may actually help the animals.

This is the fundamental problem with the emotion surrounding the animal welfare debate. As narcissistic humans, it so often becomes about us again. Our ethics, our beliefs, our tribe, our politics. Farmers take it personally – as if it's all about them – rather than taking a long hard look at what they do. They rail against terms like 'factory farming' but as the *Guardian* journalist and Nuffield Scholar Tom Levitt pointed out to me, there's no point shooting the messenger:

'Farmers need to understand why the media and public are using these terms,' he says. 'A farmer may not see their farm as a factory but if you haven't seen that sort of system before, a person may think, this looks like a factory.'

So many farmers are blinded by desensitisation. Animal rights hardliners are just as entrenched. Their minds are made up, enslaved to dogma. Blinded by sensitisation. Those that trespass on farms need to find cruelty – the crime to fit the charge – otherwise what's the point? They're hardly going to turn around and say, 'We visited a very well-run factory farm last night. Well done'.

'There are parts of the vegan community that I wouldn't go near,' says Olivia. 'I'm tired of seeing extremists ruin things for the vegan name. It's supposed to be the biggest thing of the decade and our goal is to get as many people to transition as possible. If we want to make this a thing that lasts, then we shouldn't be scaring people with extremism and horrible videos that don't really align with who we are.'

Olivia contacted me when she was working on a dual piece for the Vegan Review and her final university assignment about what it would take to get dairy farmers to transition to plant-based farming. It's an interesting and perfectly fair question – I've had the same philosophical discussion myself with farmers – yet, coming from a vegan, it bombed. Olivia ran straight into the toxic divide when she posted a poll on numerous Facebook groups.

'I got a lot of hate,' she says. 'My question was: "If you received all the help, expertise and money you need to transition, would you do it?" That triggered a lot of them. They felt I was telling them to do it, or what they were doing wasn't good enough. They just felt attacked.'

Ironically, so did Olivia.

'They trolled me. They called me bad names; things that made me feel like I was stupid, that I hadn't checked my sources. I was supposed to get 100 responses to make it a valid poll and I only got 30. I contacted as many people as I could, but I don't know many farmers. Even the National Farmers' Union wouldn't push out my poll. They see my email from the Vegan Review and they think I have another agenda. That's a problem.'

Olivia came across me through my project Just Farmers, which helps fellow journalists find independent farmer case studies. Just ordinary farmers, speaking from the heart about their own businesses and personal experiences in the industry. I was thrilled to pair her up with dairy farmer Peter Gantlett, who milks 150 crossbred cows on his organic farm in Wiltshire. His cows are a pretty mixture of Simmental, Montbeliarde, and British Friesian genetics, beautiful coppers and creams, so quite different from a traditional black and white dairy cow.

'It was so great to be able to ask Peter outright questions, just because I'm curious, and he wouldn't get triggered,' says Olivia. 'He gave straight answers and engaged in conversation and was completely honest with me. You can understand each other better when you're not putting up barriers and judging each other before you've even opened your mouth.'

I wonder what it was like for Peter? I call him to find out.

'I just explained how I farm,' he tells me over the phone. 'I talked about animal behaviour, and cow/calf separation, and how we keep the bull calves for beef. I wasn't attempting to challenge her views on veganism – I didn't want to come across that I was heckling in any way – I just put the case forward that you can farm animals in a caring way and look after them. She definitely listened and was very pleasant and I enjoyed talking to her. I know some farmers have had bad experiences on social media and been attacked but I think it was Olivia who pointed out to me that it cuts both ways; some vegans have experienced it too.'

In the end, Peter chatted to Olivia the vegan journalist for three hours. They are never going to agree in a million years, but what I find fascinating is, despite their world of difference, the London vegan and the Wiltshire dairy farmer have a lot in common. They both belong to minority groups (farmers and vegans, together, make up just 2% of the UK population) and they are interested in the same things; issues which, quite frankly and depressingly, lots of people don't give two hoots about. And they both love animals.

I'm sure this is true for most people who are deeply invested in the animal welfare debate. They're good people who care about animals, so why has it become so weirdly competitive? Warring parties take ownership of animal welfare as if they're the only ones with the capacity to care for a non-human.

It's the key reason so many farmers stubbornly refuse to engage with activists: they believe they know what's best for their animals and frankly don't value the knowledge or input of outsiders, who, they assume, wouldn't have a clue how to look after a pregnant sow, or a lame sheep, or a cow with mastitis. In the mind of a good stockman and stockwoman, everything revolves around good welfare anyway – carefully monitoring their animals' health and nutrition, watching for signs of illness or discomfort, measuring feed rations down to the last gram, making sure they have clean bedding, fresh water, and round-the-clock care during lambing, calving, and farrowing. I understand their frustration when, like banging their head against a brick wall, they point out, yet again, that their own survival as farmers depends on the welfare of their animals. Good welfare = healthy animals = happy farmers. Bad welfare = unhealthy animals = bankrupt farmers.

On the other side of the debate, activists and welfare campaigners see only the cold hand of business. Profits put before compassion; sentient beings reduced to economic commodities. And all to produce something that's killing the planet and giving us heart disease.

No, if locked in a room together, a livestock farmer and an animal rights campaigner are never going to agree on the same definition of animal welfare. Despite their shared interest in the subject, they're coming at it from opposing angles. They would interpret the same facts in a different way until the end of time.

But there is a middle ground. For increasingly ethically minded meat-eaters, and I count myself among them, simply demonstrating that healthy animals grow well and put on weight is not a convincing measure of welfare. It's not enough. Animals, just like humans, deserve a happy and fulfilled life. A good life, and a good death – and we're willing to pay more for that peace of mind. But how can we ever be certain?

The human perception of what is good animal welfare, and what constitutes animal cruelty, is constantly shifting and evolving, and varies massively around the world. In 2016, I sat next to a Chinese dairy industry delegate at a farming conference dinner in Ireland. Food safety dominated our conversation as China was still recovering from the impacts of its awful 2008 Melamine Scandal, when 300,000 babies got sick, and several died, from drinking contaminated infant formula. At the time, demand for milk powder was far outstripping supply so corrupt people in the supply chain saw the perfect opportunity to water-down raw milk in order to boost volumes. To avoid detection, a chemical called melamine was added to top up the protein content.

It caused urinary problems, kidney stones and long-lasting kidney damage in babies. The scandal engulfed some of China's biggest dairy companies and rocked consumer confidence. Considering this context my dinner companion looked bemused when I asked about dairy cow welfare. Her face said it all: 'Er, I just told you babies died from drinking milk and you're worried about the cows?' She dismissed it in a sentence, explaining animal welfare hadn't really caught on yet as food safety was still the number one priority. As far as she was concerned, it was a non-issue.

I find it fascinating when a mirror is held up to the British infatuation with animals – we're known for it around the world. It's popped up in conversations I've had, from France to Australia, and it makes me uncomfortable at times. One day on the allotment in Bristol, Conrad, our Jamaican allotment neighbour, and I are discussing politics and the Black Lives Matter movement. Lucy, our sprocker spaniel, is merrily puddling around the raised beds with a stick in her mouth. Conrad looks over to her and muses: 'In this country, the respect that they give to animals – dogs and cats – that they don't give to human beings, that they don't give to the Black people …'

He shakes his head: 'Give the people even quarter of the love that you give to the animals and the world might be a better place.'

The shifting sands of animal welfare change dramatically through place, and time. Dad remembers the last pig they ever slaughtered at Craigllwyn Farm; a large gilt and a proper character who would climb on top of the bales in the barn and make a nest in the hay, all cosy up in the rafters. On that last 'pig killing day', sometime in the 1970s, they led her from the barn and forced her to lie flat on a table as she squealed in fright. Dad, Grandad Bill, and a neighbour lay across her – it took three grown men to hold her still – while Grandad Wilfred, the local butcher, slit her throat. Nanny held a bucket to catch the blood, stirring constantly to prevent it clotting so she could make black pudding.

'I felt terrible,' Dad remembers, 'it was bloody awful.'

'Didn't anybody consider her welfare?'

'It was never even thought of. It was all about food. Animals were food.'

Food or no food, I can't imagine ever being that desensitised to an animal's terror, stress, and suffering. Surely, they knew she was feeling all these things? Where was their compassion? I feel sick just imagining it. And then I think back to those shifting sands of place and time: I've never fed a baby poisoned milk. I've never had to lie on a dying pig to get a pork dinner. My lived experience is completely different. I enjoy the privilege of compassion, in the comfort of blissful ignorance.

All the gruesome and grisly things associated with the animal products I consume have been hidden from my view. Even growing up on a farm, the 'death bit' happened somewhere else so we didn't have to think about it. As a righteous teenager, I was far more concerned about the welfare of whales and dolphins than the pig on my plate, or the lambs outside my bedroom window.

For decades, it has suited us meat-eaters not to have to look at the reality of our chosen diet, and the modern food and farming industry has been quite happy to be left alone to get on with the job behind closed doors. But turning a collective blind eye has led to a severance from the reverence. Too many of us have forgotten the basic terms of that human/animal domestic partnership I described earlier. The deal is we sustain them in life, and they sustain us in death. But it's got all blurry and mixed up. Some of the most restrictive farming systems really stretch the definition of the word 'life', limiting an animal's time on Earth to a shed and a brief and monotonous existence of eating, sleeping, breathing, and pooing. In my mind, we've become so obsessed with our side of the bargain – what we get out of it as humans – that we've forgotten that we owe something to the animals too. It's not all about us.

Meanwhile, oblivious consumers increasingly seem to discover with horror the 'death' side of the bargain, as if only just reading the small print on an ancient contract. Just as the routine sight of blood pulsing from a pig's throat desensitised Nanny and Grandad

Wilfred, and almost certainly deadened their compassion for animals, the complete disconnection from the meat production process – our separation from death – has created the opposite effect in younger generations. We have become, in contrast, overly sensitised. We no longer see food when we look at farm animals. We see cuteness, purity, vulnerability, and innocence – traits the misanthropic among us may feel are lacking in our own species. We anthropomorphise. We are overwhelmed with an all-consuming compassion. This polarisation is yet another disconnection; another symptom of the divide between those who farm and those who don't.

Animal rights groups depend on this disconnection – their narrative is weakened without it. One day in 2019, as Alex and I strolled down Broadmead in Bristol City Centre, the vegan charity Viva! had set up a street stall. A perfectly friendly campaigner wandered over with a leaflet:

'Hello there! Have you ever seen inside a factory farm?'

'Yes, quite a few.'

He looked like a goldfish for a few seconds – totally unsure what to say next.

'Oh right! Well, um, what did you think?'

'The ones I saw didn't look like that . . .'

I point towards the TV screen on their stall displaying horrific, stomach-churning images of animal cruelty. Actors are standing around it, motionless, dressed like the guy from V for Vendetta – the freedom fighter who wears a black cape and a scary white mask. It's a powerful and deeply chilling piece of performance art.

The friendly guy and I have a really interesting chat – but based on shared knowledge, not emotion. With me, they couldn't achieve the shock value they were looking for. I know too much. We could talk for days, delving into fascinating details and intricacies, but that's not what a street stall is about. The TV and the actors were there to shock, scare and sicken. It's designed to exploit the disconnect between the public and their food. The more oblivious we are, the better.

We've held onto rosy storybook images of 'Mom and Pop'

agriculture, like Old Macdonald with his one pig, one cow and one sheep. And it's a hell of a shock clicking a video on YouTube and seeing where chicken nuggets and frankfurter sausages really come from. The well-fed, deep-pocketed, non-farming population of the Western world has woken up and found farming changed – and they don't like it.

Meanwhile, farmers and the food industry completely forgot to consider how cold and cruel 'productivity and efficiency gains' might look to ordinary people. If you've been there the whole time – rolling with the changes in agriculture – maybe it doesn't seem so dramatic that you moved your pigs and cattle indoors, put them in weird contraptions, and your chickens grow faster than they used to. It's simply progress; common sense. The natural evolution of an expanding farming business. The number of farmers I meet who are genuinely shocked and offended when members of the public object to what they're doing, like they can't understand why anyone would find their practices disagreeable. 'But it's better this way!' they say repeatedly and indignantly.

In 2017 I spoke to a room full of ranchers at the South Dakota Cattlemen's Convention. One of them asked for my advice on countering anti-farming propaganda, which I get a lot. They want a silver bullet – something they can say to ungrateful consumers to make them shut up and appreciate the food on their plate. What they really mean is: 'What do we do about these dumb people who have no idea where their food comes from?' I usually offer up some advice about posting positive messages on social media but, in that moment, I decided to throw it back:

'The public didn't come along for the ride you know? They didn't see what you saw – the gradual intensification of farming, the growth of feedlots, the introduction of hormones in beef. You got used to all these things along the way, in your own time. For millions of people, now's the first they're hearing about it, and it's a shock, and they don't like it. If you don't take consumers on the journey with you, don't expect them to understand.'

A murmur of begrudging agreement rippled around the room. I believe the urban/rural divide has deepened and darkened the debate over animal welfare – a physical, geographical separation of most of the humans from most of the animals has led to a cultural estrangement. And I believe it was convenient for both sides to keep intensive livestock production hidden away like a dirty secret for years. It suited consumers not to see it and it suited farmers to be left alone to get on with it.

People I know in the UK pig industry have told me I'm wrong; that nothing was ever secret. Technically, maybe not, but it was certainly never shouted about. As a journalist it was made particularly difficult for me to gain access to intensive pig farms. I was hammering on the industry's door for years before I was ever allowed to peek inside. In the end I went to Denmark.

I was covering a story for *Countryfile* about antibiotic resistant bacteria spreading from pigs to people in 2017. So-called 'Pig MRSA'. After weeks of research and phone-bashing I eventually negotiated access on to a Danish farm which was taking part in a research project to build a better understanding of the bacteria and how it spreads. I hadn't been on many large-scale intensive indoor pig units and, yes, I admit I found it challenging. The ammonia stung my eyes, cobwebs dangled from the ceiling and I was pushed and shoved around a concrete pen by the most enormous sows I'd ever seen. Now and again a pig somewhere would let out an ear-splitting squeal like something out of Jurassic Park. We'd stop filming and wait until the noise quietened down.

I could have made that farm look like a complete hellhole. A few carefully chosen camera angles. Dim the lighting. A poignant close-up on those porcine eyes which, if you're minded towards anthropomorphism, convey all sorts of human emotions. I could have shot it to suit any agenda I wanted. And that is precisely why so many pig farmers are terrified of letting journalists and television cameras on to their farms. I was only there because the farmer had decided to trust me. I'm glad he did. His courage was a victory for transparency, and I hope I honoured it with balance in my storytelling.

The film went out on BBC One, in a primetime Sunday night teatime slot in the spring of 2017. Many viewers had never seen commercial indoor pig production before and were appalled. 'Thank goodness we don't have farms like this in the UK,' exclaimed Twitter. Some even vowed to stop buying Danish bacon.

But we have plenty of commercial indoor pig farms in the UK. I was concerned we had inadvertently led people to believe otherwise. This is incredibly dangerous. When consumers are in the dark about how their food is produced it hands the power of enlightenment to those with an agenda. If I didn't tell that story, the creepy guys in the capes and scary white masks would.

It made me even more determined to lift the cloak of secrecy shrouding British pig farms and get the BBC's cameras inside. I wanted slats, farrowing crates, docked tails – the lot. All the stuff we should know about as informed consumers.

It took me a full year to get in one. A year! I passionately made my case to the National Farmers' Union and the National Pig Association, with no luck. Their members were too scared, and it wasn't worth the risk. They were terrified of fuelling the fire – attracting more attention from activists – and, ultimately, bad news is bad for business. Far better and safer to keep a low profile. They weren't interested.

In the end I found a pig farmer through my own network of contacts – but it still took six months to persuade her. I drove back and forth between Bristol and her home in the east of England, reassuring her over cups of tea that it was the right thing to do, that she would be doing a service to the public and her own industry.

In March 2018, the doors to a British intensive indoor pig unit, complete with slats, farrowing crates and docked tails, were finally opened to our cameras. We decided not to name the farmer, for her own protection, but she showed her face and answered every question and challenge our presenter Tom Heap threw at her. We did not give her an easy ride, but she took it head on.

She didn't regret it either. I warned her to stay clear of social media on the night the film was broadcast but she later told me she 'couldn't resist a peek'. There was a lot of hate, which she expected, but the reality wasn't as bad as the fear she'd built up in her mind.

Now, several years later, the industry has opened up a lot. I've been on dozens of intensive pig, dairy, and poultry farms. It doesn't feel like such a big deal anymore, and rightly so. We should be allowed to see where our meat, milk and eggs are produced. And these are legal farms producing food that millions of people buy – why should they skulk in the shadows, hiding from the rest of society?

So, after my own journey through the animal welfare rabbit hole, have I found a way out? Well, I've tunnelled a passage that works for me. When I can, I choose to buy outdoor-reared pork, grass-fed beef, free-range chicken and organic milk and eggs. I do prefer to see animals outside. I don't have a problem with indoor livestock production per se (if I were a cow, I might even prefer to spend the winter in a nice warm shed) but it's all those conflicts and contradictions that mess with my head. We seem to make life much harder for ourselves, and for the animals, locked in a perpetual game of welfare whack-a-mole. It's like journalism and storytelling – keep it simple, stupid. I can't help but feel the same about farming. Let a cow graze. Let a pig root. Let a chicken scratch and peck at the earth.

But the key word here is 'choose'. It costs a lot more to eat what I perceive as a high-welfare diet, something I've only been doing in recent years since living with Alex and having someone to share bills with. When I rented flats in the city on my own, I was cash-strapped and constantly in my overdraft. I bought four pints of milk for a pound and conventional meat from the supermarket. Sure, I could have afforded 'better food' if I'd forgone afterwork drinks, meals out and cooked everything from scratch, but why would I do that? I was a busy young professional enjoying my life and supermarket food was perfectly good enough. I could never campaign against the existence of intensive livestock production because I would very quickly be exposed as a hypocrite, and it only takes a small change in my circumstances for me to be digging around in the Aldi refrigerators again.

Until we all decide to go vegan, or we figure out a way of making organic and regeneratively farmed meat and dairy the new bog-standard baseline on the cheap shelf in the supermarket, we need conventional production systems.

Intensive livestock farming isn't going anywhere fast, however much we shout and scream and rant and holler. We are stuck with it for a while longer and I, for one, would rather accept that fact and try and make it better now, today, than fixating on some utopian plant-based future.

So, what do we do?

I think I've seen intensive indoor livestock production done as well as it can be. I've been on dozens of so-called 'factory farms'. It's not pretty or pleasant – my hair and clothes stink afterwards – but it's not the perpetual horror show some activists would have you believe. I've donned the bio-secure overalls on intensive poultry units and walked among chicks, young broilers and laying hens (both 'enriched colony' caged and free-range). The birds looked healthy, well cared for and the sheds were clean and provided limitless water and some 'enrichment' (perches and things for the birds to climb on). I was not repulsed. However, I've never been permitted to see fully grown broilers at the point when they're ready to leave the farm. This lack of transparency at the final stage in the production system does concern me. It feels off-limits.

On the best intensive pig farms I've seen, as far as I could tell, the animals weren't hungry or thirsty, they didn't look uncomfortable, diseased or in pain, and they weren't visibly distressed. If normal behaviour includes nibbling at your wellies and being very nosy, they did plenty of that. I do believe they're bored, particularly young energetic pigs in slatted units where there's no straw to root around in. I struggle with that. That's my own personal impression. I'd urge you to try and see it for yourself and make up your own mind – and not by creeping around some farmyard in the dark with a camera. Do the decent thing and reach out to a farmer. Build a relationship. Ask them to show you around. Leave pre-conceived ideas of what you think a farm should look like at the door and try and accept it for what it is.

However. Those are the good farms. There are bad ones. Animal rights activists have exposed some truly awful places – genuine houses of horror. It's not all made-up 'anti-farming propaganda' as the industry sometimes likes to paint it. There are farmers and farm workers, moving in our circles, who bring shame upon their vocation. I haven't met many in my lifetime, and thankfully none (that I know of) in recent years, but I'm sure other farmers, farm workers, children of farming families, and people who work in livestock markets and abattoirs have also crossed paths with at least one person who really shouldn't be working with animals. They are out there, and it angers me when we pretend they're not; when industry voices say things like it's just 'one rotten apple in the cart' and 'every farmer puts the welfare of their animals first.' It simply isn't true. I find the lack of accountability I sometimes encounter maddening – how close-knit farming communities shun these cruel bastards privately, but close ranks and keep quiet publicly, perhaps believing in some twisted loyalty to country values. I call on the farming world to turn on your rotten apples. Stop protecting them. The damage they are doing to our farming industry is a thousand times the damage wrought by animal rights activists.

I must also stress that bad animal welfare is not limited to 'factory farms', it can just as easily happen on smallholdings, but large-scale industrial units have the potential to cause much greater harm for the simple fact there are hundreds and thousands of lives in their care. Farmers have to be at the very top of their game to run these systems well – the risks and responsibilities are enormous. Get it wrong and you have a house of horrors on your hands. One bad day, one rogue member of staff, one video camera is all it takes – and it'll drag the rest of the industry down too.

There's no way I could handle that kind of pressure – I respect those that do – but as a shopper I would rather not take the risk if I can avoid it and afford it. That is my personal choice.

So, here I am emerging from the welfare and animal rights rabbit hole into a beautiful meadow grazed by gorgeous grass-fed cows. I'm blinking in the sunlight, breathing in the fresh air of a decision made: I'm comfortable with the ethics of what I'm eating. Good job. But suddenly... bang. A badger drops dead in front of me, shot as part of the cull to control the spread of bovine TB in cattle. What happens when a farmed animal's welfare is pitched against the welfare of our native wildlife?

And I'm back in the rabbit hole again.

CHAPTER SIX

FOOD

W E ALL HAVE TO EAT REGARDLESS of whether we live on a croft in the Scottish Highlands or in a flat in Tower Hamlets. Surely food is the umbilical cord connecting everyone with the land?

Nope.

Food and farming are often mentioned in the same sentence. Every year the BBC hosts its celebrated Food and Farming Awards where chefs share the stage with farmers and growers. It's a beautiful image to see these two worlds presented as so intertwined: like players on the same team lifting a trophy together. But it's true only for certain types of farms and certain types of chefs. Some farmers, the minority, sell direct to restaurants, butchers and bakeries and I'm sure some food retailers, the minority, are on first name terms with their farmer suppliers and exchange Christmas cards. The UK's 1,000 farm shops are a cosy shop window for British agriculture and give the warming impression of the interconnectedness of food and farming. It's real, for sure, but it's not the whole picture, far from it. In fact, food and farming can be worlds apart, and the urban/rural divide is one reason for the gulf.

A connection to food does not automatically mean a connection to farming. I'm yet to meet anyone else from a rural and agricultural background in my Bristol inner-city neighbourhood. From my house I can see terraced houses, cars, scaffolding, TV aerials and telegraph poles. We don't have a garden – just a small backyard paved with concrete slabs, a shed for our bikes and some rather sad-looking potted plants which Lucy nibbled as a puppy. Farming and the countryside feel very far away – yet food culture in this deeply urban area is flourishing. There's a wholefood deli and artisan bakery just a two-minute walk up the street. I can explore the cuisines of India, Poland, Somalia, China, Thailand, Italy, Latin America and the Caribbean all within a square mile of my house. I can eat organic, free range, vegetarian, vegan, or gluten-free every day of the week and without much effort. You don't need fields and farms to find good food suiting every taste and diet.

For me and most of the 67 million people living in the UK today, our main connection to farming is through the food we choose to buy in the shops. The food that fits with our wallet, our tastebuds and, if we're lucky enough to afford it, the food that fits with our ethics.

What we choose to eat may help us develop an opinion about farming, but a connection to farming? Not really.

You can love food and call yourself a 'foodie' and know absolutely nothing about, or have little interest in, how it was produced. I use my partner Alex as the perfect example. He absolutely adores a street food market. Alex's happy place is chowing down on a £10 burger grilled in a converted VW camper van by a tattooed hipster with a beard. Chuck in a craft ale pulled out of a stainless-steel vat from a microbrewery on an industrial estate and he's in heaven. Jerk chicken, pulled pork, wood-fired pizzas, ribs, pork belly, beef brisket – if it's a 'food experience' he'll pay whatever price is written on the chalk board, no questions asked. He could barely contain his excitement when a graffitied van pitched up in Bristol selling street food. Alex and his mates sat cross-legged on the ground, in the car park of a roof tile factory, eating 'Grizzly Burgers' smothered with oak-smoked bacon and sticky bacon jam on brioche buns. He remembers the ingredients, even put it on Instagram, but at no point did he ask where the beef was from, wonder how it was reared or give a moment's conscious thought to the fact his beef burger was once a cow, on a farm. Bears were probably closer to mind than cattle.

Alex wasn't thinking about 'food and farming'; he was thinking about food. In fact, before we met, farming never crossed his mind. He didn't read labels on packets of meat in the supermarket, or even notice whether it was free range, organic or Red Tractor. He's never asked a restaurant where their ingredients are sourced from.

Alex, like millions of consumers, values taste and affordability. He'll shop on price for the weekly staples and spend more, a lot more, on treats or 'food experiences', like his Grizzly Burger. I am not judging him, nor am I some kind of ethical food archangel sent to show him the error of his ways and set him on the righteous path to food provenance. I am in no place to judge. For all my pontificating to poor Alex about buying British free- range chicken and outdoor-reared pork, I don't ask our local curry house where the chicken is from in my Friday-night dhansak, and I won't let my reservations about intensive, indoor pig production stand between me and a pepperoni pizza. I live in a giant glasshouse wielding a massive rock. Like most perfectly normal, well-meaning consumers – I am a big fat hypocrite.

Food is not the conjoined twin of farming. For most consumers food is a best friend we see every day and farming is a distant, long- lost cousin we haven't heard from in years.

This causes distress and consternation in farming circles. I can say with complete confidence, and without the slightest exaggeration, that the issue raises its head at every agricultural conference and trade event I've ever attended: the Great Disconnect between the masses and their food. In Q&As and PowerPoint presentations, during conference buffets and coffee breaks, in Zoom chats and webinars, someone will always ask about bridging the gap between farmers and that shadowy group of people 'who don't know where their food comes from'. Many simply assume it must be 'poor people' who eat in such blind ignorance.

Not necessarily.

'The people who are not caring about where their food comes from are wealthy,' says Essex farmer George Young, 'They are the people who value spending £80 on the gym every month and have committed to eating less meat for environmental reasons, but when they eat meat once a week, they'll say they can't afford to spend more money on that meat.'

George brings a unique perspective from his years spent hopping between urban and rural life. A fourth-generation farmer's son, he was brought up on his family's 1,200-acre arable farm in the village of Fobbing, just 30 miles or so from Marble Arch in London. As a teenager, he harboured no ambitions to follow his father into farming so focused on becoming a musician instead, playing clarinet and saxophone to a semi-professional level. He moved to London and slotted gigs around a stressful career in banking. George lived a fast life in the capital and barely slept through most of his twenties.

Insomnia is about all he took from London life when he returned home to the farm in 2013. In the process of reinventing himself from financier to farmer, he's set about reinventing the family business from a large-scale conventional arable operation to a blueprint for agroecology. He diversified the crops, introducing buckwheat, hemp, flax, lentils, teff and ancient wheat varieties to grow alongside the old reliables of commercial wheat and peas. He bought a herd of red poll cattle, so he can raise and market his own pasture-fed beef. He developed a passion, bordering on obsession, with farm-to-plate principles, even learning how to mill flour from his own grain. He imported a traditional stone mill from Vermont in the US and installed it in an old grain store in the farmyard. He sells around 50 tonnes of milled wheat, as flour, to artisan bakeries and sells grain in their raw form to Hodmedods, a company which specialises in sourcing and marketing British grown pulses and 'alternative' grains.

The busy city boy who barely found time to eat has transformed into a one-man embodiment of 'food and farming'. He relishes the joyful simplicity of making food from what he can grow on the farm and selling it to people who appreciate it. He marvels at how uncomplicated the transaction is. Compared to his commodity crops, which disappear up long and complicated supply chains like a giant beanstalk reaching from the Fobbing soil to a faceless supermarket in the sky, this is the stuff of primary school maths. George receives roughly £200 a tonne for his conventional wheat (depending on the market) while the niche crops can fetch up to £650. Sure, the yield is just two-fifths of his commercial crops, but they're grown with far fewer inputs (fertilisers and herbicide).

'I'm not making loads more, but it's better than the poorly paid system which preceded it,' he says.

George felt reconnected to something deep and primal and set out to share his epiphany – that paying a little bit more to know where your food comes from is an act of reconnection to nature, to the land and to your own health and wellbeing. He started off with his wealthy London friends, the ones who said they couldn't afford to spend more on meat:

'I always pull them up on the use of the word afford,' he says. 'If you're earning over £25,000 then you can't use the word afford. The word they're looking for is value'. They don't value food. But they value going to the gym.'

It didn't go quite the way he hoped. Even when they saw George's cows grazing in the sunshine, and commented on how beautiful they look, they still wouldn't commit to spending more on meat:

'It makes me really frustrated,' he shakes his head. 'I assume it is the dominant perception. When you look at the village here, an affluent commuter village, I'm pretty sure that is the prevailing attitude. People have been indoctrinated to think that price is all that matters.'

Bubble burst, George ran slap bang into the same downhearted gloom felt by many farmers when they ask themselves: 'Why don't people value what keeps them alive? Why don't they value my labour in producing it?'

Soaring food prices and the cost-of-living crisis, which engulfed Britain following the Russian invasion of Ukraine in February 2022, has piled greater pressure on consumers. Now, rather than simply choosing not to spend more money on food, many shoppers (myself included) are discovering with horror the alien experience of parting with more cash for less stuff. People my age and younger have never known food inflation on such a scale – things I expect to be cheap, are no longer cheap. I find myself tutting at price tags on supermarket shelves (five pounds for a box of teabags?!) I loiter in the aisles, picking things up and putting them back again. I used to put chewing gum in supermarket receipts, now I check every item in case they've missed off a multibuy saving. The fact is I am simply not used to spending a greater proportion of my income on groceries. I am a child of the cheap food era. In the 1950s, we spent a whopping third of our household budget on food. By 2020, it had dropped to roughly 8%. Shifting our spending away from food has often been pounced upon by frustrated farmers seeking to confirm their suspicions: 'See? They used to care about food but now they'd rather spend money on iPhones and PlayStations!'

Even with food prices rising at their fastest rate in 45 years, I heard a Welsh dairy farmer grumble: 'They say no one can afford food but the football stadiums and nightclubs are still full every week. They're finding the money from somewhere!'

As a general rule, when incomes rise, the proportion of income spent on food falls. Even if everyone bought more expensive food and switched permanently from Lidl to Waitrose, ultimately, we can only eat so much. Our demand for food, like our waistlines, can only stretch so far, whereas demand for PlayStations, season tickets and going clubbing can keep growing and growing.

Still, the relegation of food down the list of household expenditure has contributed to the unshakeable feeling that farming has been relegated too. It has led to a collapse in self-confidence and a feeling of irrelevance among many farmers and growers I speak to. They'll take to heart a one-off survey that reveals a small number of schoolchildren think pasta grows on trees or pigs produce fishfingers. They'll hold it up despairingly as the final depressing piece of evidence confirming their worst fear: urban people now believe their food comes from a supermarket.

My hunch is the perception is worse than the reality. The vast majority of kids know milk comes from cows, if only because there's usually a picture on the bottle. Either way, this feeling of being out of sight and out of mind has contributed to the divide between rural communities who grow the food and urban communities who eat the food.

But it works the other way too. This disconnect between food and farming is not, as many in the agricultural industry lament, a symptom of urbanisation – a metropolitan disease. A connection to farming does not automatically mean a connection to food.

You can farm your entire life and care little about what happens to your crops or animals after they leave the farm. Making your living from commodity production, whether it's growing cereal crops, fruit and vegetables or rearing animals, does not necessarily translate into an appreciation for, or any interest in, how your products are processed, cooked, and eaten.

George Young says many of his farming colleagues and peers have zero interest in food. He'll ask them about it at local NFU branch meetings:

'I ask arable farmers about where their crops go – most of them have no idea. 'Where do your potatoes go?' I say.

'Chips I think.'

'Where does your wheat go?' 'I think it goes to Warburtons.'

'How do they make the bread?' 'No idea.'

George throws his hands up in exasperation: 'They have pride in what they grow but don't give a shit about what happens to it after they've sold it to whoever is going to bastardise what they've grown! I find it perplexing how little they care.'

From my experience, many commercial farmers, Dad included, are far more motivated by national food patriotism – the notion of British food and self-sufficiency – than short, local supply chains and selling direct to their consumers. Their pride comes from feeding the nation, filling the supermarket shelves, and keeping millions of bellies full. How we fiddle about with that food in factories, supermarkets, restaurants, cafés, and kitchens is not their concern – so long as it's British. This is without doubt a legacy from the hungry post-war years when food was about keeping you alive, not for posting on Instagram.

'Food is fuel,' Dad would tell us as kids, a sentiment that sums up the attitude to calories I've witnessed in many rural communities. I grew up with a practical, utilitarian relationship with food – it gave us energy to grow, to work. Like filling a tractor with diesel. The stories and ceremony around food, that Alex will pay through the nose for in Bristol, rarely featured in my childhood.

I don't remember eating our own livestock – Mum says we always had our own beef and lamb in the freezer, but I don't remember. It wasn't celebrated. I've eaten cheap processed meat of unknown origin in the canteens of many livestock markets. So has Dad.

'They used to sell some awful tack,' he remembers. 'Perfectly square bits of beef, the same shape as bread, and when you held it up to the light you could see straight through it. I used to play holy hell with them — we're farmers and you're making us eat this rubbish!'

Some of the working farmhouse kitchens I knew best as a child served white sliced bread, still in the plastic bag, plonked in the middle of the table. We'd dig in, pull out a slice and butter it ourselves unceremoniously. Hardly the homemade rustic loaves in wicker baskets shrouded in gingham cloths envisaged in romantic depictions of rural life (or set-dressed on *Countryfile* during every picnic scene I've ever filmed).

Our food may not have been gourmet, local, seasonal, or organic — but it was always stamped with the Union Jack. I'd cringe with embarrassment when Dad held up the queue in McDonald's demanding to know if his Big Mac was British beef. He always seemed braced for catastrophe, like British food was a fragile thing, easily lost and to be defended at all costs. I never really understood that strength of feeling and confess to even rolling my eyes when farmers got on their soapbox about 'buying British'.

But I get it now.

In 1984, when I was a toddler, the UK was 78% self-sufficient in food. Today, it's fallen to around 60%, some of which is exported. We import 46% of the food we consume domestically — nearly half — and I can only see that figure rising if costs of production keep overtaking the prices that farmers get paid, and the Government gives increasing access to our food market to some of the world's largest agricultural exporters.

I've been covering the ups and downs in British agriculture for years, reporting on various crises in different sectors. It'll be dairy one year, cereals the next, pigs the next, and on and on it goes like a merry-go-round. I've come to expect the usual prophecies of doom from farm unions and trade associations warning of certain industry collapse. And then the markets pick up, prices increase, and everything goes suspiciously quiet. No farmer likes to admit when they're making money.

This constant rollercoaster – or 'market volatility' – makes headlines in the trade press, or on Radio 4 if you're up early enough for *Farming Today*, but it's undeniably niche in 'normal' times. Farmer drama has never been high up on the mainstream news agenda. As one news editor once said to me: 'Look, we have more stories than we know what to do with…it's not relevant to write about [farming] for the sake of writing about it. Unless it's got immediate relevance to our readers then why would they be interested?'

'They [farmers] have a chip on their shoulder about not being appreciated,' they added dismissively.

And that's how it was, certainly for most of my time as an agricultural journalist. For years, very few people outside the farming bubble gave two hoots about the price of milk, or eggs, or pork. We even used to joke about it in the *Countryfile* office, 'Farmers expect us to report the primestock prices – but who would want to watch that? Boring!'

I suspect that's how the retailers preferred it too. Let the commodity market deal with its ups and downs but keep the public in the dark. If farmers and growers are having a tough time on the ground, never EVER let the shopper see empty shelves in the supermarket. That would be unthinkable.

Until now.

As we hurtled from one shock to another – Brexit, a global pandemic, war in Ukraine – the cost of running a UK farm exploded: workers' wages, gas and electricity, machinery parts, animal feed, fertiliser, diesel for tractors, even the small stuff like iodine to dress the navels of newborn lambs as much as quadrupled in price. In 2022, the cost of producing food soared beyond anything I'd seen before. Sector after sector plunged into crisis. It felt like several years' worth of farming headlines were hitting the front pages every week.

Predictably, industry groups cranked up the doom machine, warning of food shortages unless the prices paid to farmers increased to cover their costs of production – eggs in particular.

The supermarkets were painfully slow to react. Exactly why depends on your perspective. You could argue they had to think of increasingly hard-up customers and keep prices as low as possible in a cost-of-living crisis. One industry source even hinted to me that some farm organisations have a reputation for 'crying wolf'. Perhaps the retailers didn't believe threats of shortages were sincere – just more doom-mongering from the whingeing farmers.

An alternative view is that a toxic culture of squeezing farmer suppliers had blinded buyers to the seriousness of the situation on the ground. Rumours of 'Greedflation' took hold, accusing big food brands and retailers of profiteering from inflation.

Whatever happened they were caught off guard, the supply chain buckled and, just as farmers had warned, shops ran out of eggs.

In November 2022, some supermarkets blamed Bird Flu for the shortage because more than half a million laying hens had been culled due to the disease. The UK had never seen such a serious outbreak.

But many poultry farmers refused to buy this as an excuse. One of them was Welsh egg producer Ioan Humphreys, a member of my Just Farmers network, who took to social media to vent his frustration. Standing in front of pallets piled high with free-range eggs, he spoke angrily to the camera and posted on Instagram: 'The current egg shortage isn't really to do with Avian Flu, it's because the supermarkets are refusing to pay us a fair price for what we produce.'

Ioan laid it all out there. He explained how the cost of new hens, to restock the sheds after the previous flock had reached the end of their laying life, had gone up by £1 per bird. He runs a 32,000-bird unit. Those chickens peck through 30 tonnes of chicken feed a week. His feed price had leapt from £250 to £400 a tonne.

The average cost of production, at that time, was anywhere between £1.20 and £1.40 per dozen eggs. Ioan was getting paid £1.09.

Meanwhile, shoppers were paying around 50p more for eggs at the checkout. Farmers knew there was more money coming in at the top so why wasn't it filtering down to them at the bottom?

And that's when some poultry farmers effectively stuck two fingers up at the system and decided to hold off on restocking their sheds. Coupled with the losses from Bird Flu, the drop in egg production led to empty supermarket shelves.

Farmers aren't muppets. They'd crunched the numbers, shook their heads, and those that could take the hit, decided an empty shed made more financial sense than producing eggs.

Since then, the egg price has recovered – by quite some margin for some – and Ioan seems happier.

But just as one sector picks up, another sinks into crisis.

Across the UK, apple growers are doing exactly what the egg producers did – not replanting crops or grubbing up entire orchards because the business of growing British apples has become totally unviable. The cost of producing an apple – planting, growing, picking, storing, hauling – increased by 23% in the space of one year. Yet growers received a piddly 0.8% price increase.

I can keep going.

British pig farmers, particularly independent producers, have been through an unbelievable period of misery. In December 2021, more than 200,000 pigs – either ready for slaughter or needing to be moved – were stuck on farms because of staff shortages in abattoirs. Thanks to the Government's ideological aversion to 'unskilled foreign labour', ironically, there weren't enough butchers, foreign or otherwise, with the necessary skills to cut up the carcasses. So, despite processors being contractually obliged to collect the pigs, the lorries simply didn't turn up. They just left them there for the farmers to deal with. The backlog, which dragged on for months, meant farmers had to manage overcrowded farms, find room for near-worthless pigs as prices collapsed, and manage a constant stream of new piglets. You can't just turn the tap off in livestock farming.

In March 2022, compounding an already desperate situation, the cost of feeding those near-worthless pigs became truly crippling, as wheat prices soared following Vladimir Putin's invasion of Ukraine.

By the end of 2022, struggling producers were selling their breeding sows (mum pigs) to reduce the number of piglets being born and save money on feed bills. The breeding herd fell to its lowest level in two decades, and the industry started losing people, as well as pigs.

I want to check in with Robert Lasseter, the pig farmer with the 360 farrowing pens in Dorset to see how he's doing.

'We are stopping pig production and we will have no pigs here from November 2023,' he says matter-of-factly over Zoom. I'm not expecting him to come straight out with it. Robert is a pig farmer. He's always been a pig farmer. Farmers like him don't just give up. It's May 2023 when we speak and the pig price has recovered, to record levels in fact. Surely, he's making money again? Somehow, it's not going in.

'But. . .why sell up now? Why not stick with it?'

'Anna, we've lost hundreds of thousands of pounds in the last two years and that hole will never be filled. We've ended up with more debt that I haven't got a chance of getting back. The risk of that happening again is just too great. I know loads and loads of people who are getting out.'

'But. . .what are you going to do about your debt?'

'No idea. But I know the pigs aren't going to solve it.'

He looks back at me on the webcam. He's unnervingly calm and pragmatic.

'The industry is going through massive structural change,' he explains. 'I was told the other day that 32 farmers, including the large processors, own 80% of the UK pig herd. As a small family business, you just think, "this is ridiculous. I'm not going to buck the trend anymore." This isn't about farming – this is about the industrialisation of food.'

He almost shrugs.

'But. . .don't you care?'

'I care deeply but I can't do anything about it. We've got no clout at all. There's no point stamping our feet and saying it's not fair when you're working with multinationals. They're in the big boys' league and

I'm in the family farmer league. It's business and I'm out of the game.'

'How can you have such a cheerful voice of surrender?' I ask, perhaps naively.

'I think it's the cheerful voice of reality,' he replies. 'Our government is leading us down the road of an internationally supplied diet. They are not interested in agriculture. We're a complete irrelevance.'

Against this backdrop of crisis in multiple farming sectors, the UK government has been busy signing new free trade deals around the world. The UK (or London at least) has some seriously big fish to sell – namely a financial services sector worth billions which represents a colossal 8.3% of our overall economy.

Agriculture is less of a headliner. It makes up just 0.5% of our national income (GDP). When you lump farming in with food processing it becomes the oft-quoted 'largest manufacturing sector' but, in truth, farming on its own is small fry compared to the whale of finance.

But that's not the case in other countries. In New Zealand, for example, agriculture accounts for around 7% of GDP – nearly as much as our financial services sector. It exports eye-popping quantities of food, and more dairy products and lamb than any other country.

With our large domestic population and comparatively small farming industry, the UK is an attractive market for food-producing giants like New Zealand. And that makes agriculture a tempting bargaining chip for UK negotiators looking to sell other services, like London's money stuff.

In May 2023, new free trade agreements with New Zealand and Australia came into force. Both were heavily criticised by the National Farmers' Union for eliminating tariffs for agricultural products. Ultimately, there will be no limit on the amount of beef, lamb, dairy, fruit, or vegetables that Australia and New Zealand can export to the UK. All things we're very good at growing ourselves.

But that's the nature of free trade, right? It's all about give and take.

I'm not anti-trade, nor a blinkered food nationalist. I think it's incredibly arrogant to proclaim British food as 'the best in the world' when the world is full of fantastic farmers, producing delicious, safe, and nutritious food for their people (many of whom are a hell of a lot healthier than us). I'm grateful to live in a country wealthy enough to import affordable food we can't grow ourselves and plug seasonal gaps in supply. Allowing access to our food market can even help lift some of the world's poorest farmers out of poverty, it gives us a seat at the table to influence farming practices and share our food and environmental standards on the international stage. Some deals may even hold some juicy export opportunities for British food. So, I am not a protectionist.

Where I do have a problem is when free trade becomes unfair trade; the risk of importing food produced on farms that aren't required to obey the same rules as our own farmers. These rules were put in place for good reason – to protect our environment, our health, or improve the welfare of farm animals.

For instance, an Australian arable farmer can use the highly toxic herbicide paraquat to control weeds in a crop of wheat. Paraquat has been banned in the UK since 2007.

Australian, American, and even some New Zealand beef producers can implant cattle with growth hormones to make them fatten faster. The use of hormones in British beef production was banned in 1989.

And here's the rub – being permitted to farm without so many rules generally saves a farmer money, meaning they can produce cheaper food. And what do shoppers love more than anything? A good bargain.

I have eaten hormone beef. It was a certified Angus steak grilled medium rare on a barbecue overlooking the South Dakotan prairie at sunset. Honestly? It was one of the best steaks I've ever tasted.

As a British beef farmer's daughter, I was worried. Because I know that steak would sell itself on a supermarket shelf – hormones or no hormones. This is why the fight to protect British food standards in free trade deals was described by NFU president Minette Batters as 'the biggest political battle the union has ever faced'.

Are they winning that battle? Well, here's one more example.

In 2012, the EU banned barren battery cages for laying hens. They were small wire cages, with each hen having less space than a piece of A4 paper to move around on. They were replaced by larger 'enriched colony' cages.

In 2016, every major UK supermarket went a step further by committing to going completely cage-free by 2025. Responding to these signals from government and retailers, many poultry farmers invested millions into converting their farms, ditching cages altogether and moving to the barn system (eggs from indoor hens) or free range.

Fast forward to 2023, the UK joins the catchily-named Comprehensive and Progressive Agreement for Trans-Pacific Partnership (CPTPP). It's a trading bloc of 11 diverse countries, from Malaysia to Mexico, Chile to Canada. As part of the deal, the UK agreed to phase out all import tariffs on Mexican eggs.

In Mexico, 99% of eggs are laid by hens in battery cages.

Unsurprisingly, the British egg industry flipped its lid and begged the government to consider the morality of opening our doors to eggs that would be illegal to produce in the UK. It was shrugged off as unlikely to happen anyway, since Mexico prefers to keep hold of its eggs to feed its own population (who love eggs, consuming around 400 per person, per year – more than any other country).

But here's what's got the British egg guys ruffled: if Mexico ever did decide to send us some eggs, it's unlikely they'd appear on the shelf next to your box of British organic. No, a battery cage egg is far more likely to be imported in powdered form and get mixed into biscuits, mayonnaise, instant custard, or some other processed product. As far as I see it, the real battle is for the hidden ingredients in your cake from the corner shop.

'However smart Minette Batters is, she's on a complete losing wicket,' says Robert, the soon-to-be-ex-pig-farmer. 'This is going nowhere but us ending up with cheap food from wherever it comes from in the world. We may not like it as food producers but, frankly, I'm coming to the conclusion we may as well embrace it. That means as a farmer I've got to find other income streams. So, we're looking at solar, developing buildings, all this biodiversity net gain and nutrient neutrality stuff. And we're also looking at being very poor.'

Personally, what I find most worrying is not the economics (farming has always been an up and down industry), or the trade deals (which aren't all bad) or the politics (which will change), but the mood of the industry: how quickly farmers have become demoralised and disincentivised to produce, as if this really is the final straw. An NFU survey in May 2023 found farmer confidence is at its lowest since the start of the pandemic. For the first time in my life, I am worried for the future of British food. I fear we're seeing the decline of our farming community.

Yet what do I do?

I still walk through the doors of the supermarket. I still push a trolley around. I still get points on my loyalty card. I'm feeding the system.

During the pandemic, I thought I'd changed. Alex and I gave up going to supermarkets and supported our local bakery and wholefood shop – which cost more but meant we became obsessive about food waste. Not a morsel was thrown away. We started buying all our meat either from farm shops or direct from farms. We took time to ask questions and even spent an afternoon squelching through Gloucestershire fields in the rain seeing first-hand where our Pasture for Life certified beef box came from.

And then the world changed again.

Long working weeks returned. The busy weekends. The need for convenience. Bills went up…and up. Money started disappearing like sand through my fingers. Supermarket bread and plastic trays of mince gradually replaced my sourdough loaves and beef boxes.

The truth is, after all my preaching, I can't wholly escape the supermarket or the conventional food system. I need it. And most ordinary folk, including farming families, are the same. I remember the hill farmer's words: 'I'm a victim of it but I'm also a perpetrator of it. I unfeather my own nest.'

But that doesn't mean I'm powerless. If it's in season and we can produce it here in the UK, it has to be British food in my basket. I scrutinise every packet for country-of-origin labelling and if I can't see the Union Jack on the shelf, I'll ask for it. If it's more expensive, I'll pay it. If it's not there, I'll walk away. For me, as a resident of these tiny isles in the North Sea, buying local means buying British. I'm sending a signal to the supermarkets that homegrown food is important to me.

I do ask myself sometimes, 'Why do I care? Should I care? Am I out of touch? Backward looking, and sentimental?'

But when I weigh it all up in my mind, it just makes sense – socially and environmentally – to eat what we can grow. Our land and climate can produce a staggering abundance of food: fruits, vegetables, salads, herbs, nuts, eggs, milk, cheese, pork, beef, lamb, chicken, turkey, even sugar. If I can get an apple from Herefordshire, why would I buy one from Argentina? I shall buy Malbec from Argentina.

I live in a country that can feed its people, mostly from what it produces itself, and what a humbling privilege that is. We are so lucky to know what food security feels like – but maybe Dad was right to be on his guard. If we lose it, it's not easy to get it back.

Ultimately, the biggest battle farmers face is not with the government, or the supermarkets, but the millions of consumers whose purchasing power rules our food system. How to instill the pride I feel, as a farmer's daughter, into their shopping habits? Does it even matter?

'No, it really doesn't,' says Robert. 'You talk about Britishness but I'm afraid it's absolutely irrelevant.'

Essex farmer George Young can't help but feel sceptical too:

'I see farmers using #FeedtheWorld like they're trying to instil their own pride into something other people don't have pride in. They keep saying people value Britishness. I'm really sorry, but many of my friends in London just don't. They are educated and they understand food, but they don't value Britishness – they value price. They like going out for a nice meal but they wouldn't choose a restaurant based on where they source their food.'

So, the notion of food and farming as cosy allies, latticed together like homemade pastry, is often a romantic ideal, a fiction that makes us feel closer to where our food comes from. The simple fact that farms in rural areas produce food and people in urban areas eat food is not, on its own, enough to bridge the divide. In fact, food choices and dietary identities in recent years have been incredibly toxic for urban/rural relations. I'm talking, of course, about the vegan debate. Meat versus Plants is a familiar battleground in the great urban/rural divide.

From a rural perspective, veganism and vegetarianism are traditionally associated with 'townies'. Rejecting meat instinctively feels like a predominantly urban lifestyle choice and I'm interested to discover it bears out in the statistics. According to The Vegan Society, 88% of vegans live in urban or suburban areas, like young London journalist Olivia Rafferty.

She was born and raised in Milan, the daughter of British and Irish teachers, and moved to London in 2018 to study Journalism at City University.

Olivia embraces the vegan diet wholeheartedly and cooks most things from scratch, replacing her previous pescatarian diet with chickpeas, spinach, tofu and plant-based 'milks'. Perhaps none of this is surprising for a 20-year-old Gen Z but get this – she even converted her parents to veganism:

'Dad was a huge fish eater and Mum was a cheese lover, so it wasn't easy for them. When I went home for Christmas, I cooked every day for everyone, so it was always vegan. If it tastes good, they will eat it. Mum struggled with it a bit at first but as far as I know she hasn't eaten any cheese. And Dad has always had psoriasis but it cleared up after a month going vegan for Veganuary.'

Olivia giggles at my open-mouthed amazement over Zoom. You're more likely to catch Gwyneth Paltrow in Burger King than see my Dad ever go vegan.

Living on a farm, seeing the cyclical journey of the animals from birth to death, year after year, you grow up with the innate knowledge that animals die so we can eat them. That's just how it is.

I encountered my first vegetarian in primary school, around the age of nine or 10. A new family moved to the area from Manchester, taking on a rundown farmhouse. One of the daughters was my age, and vegetarian, and we didn't get on at first. We had terrible quarrels about meat-eating in the school hall during lunch break. After school, like a tired boxer at the end of a round, I'd go home feeling bruised and downhearted and talk to Dad who, like a coach at the ringside, filled my ears with more ammunition for the following day. He seemed to take it as a personal affront that there were vegetarians in the village school.

In the end, we got tired of arguing and realised we rather liked each other. We became friends and remained close all the way through high school and sixth form. I even became a vegetarian for a few days around the age of 13, until Dad noticed and I caved immediately, secretly pleased to be able to eat beef again.

Why did I find my schoolfriend's personal dietary choices so confronting? Why did I feel the need to fight her? Because it threatened our identity. Someone thought our way of life was wrong, which suggested there was something wrong with us. It rocked my foundations and it scared me. I remember the desperation I felt during those arguments, the need to win. Not because I was some nine-year-old sadist who relished killing animals and gorging on their flesh, but because I needed her validation. I wanted her to say, 'OK, I see now.' I wanted the natural order to be restored. I wanted to go back to how it was before I knew a vegetarian. Aged nine, and I was already nostalgic for the status quo.

Now, I'm so grateful for those arguments. I'm glad I was challenged at such a formative age. It opened my eyes to other ways of eating – and thinking – and no doubt helped me grow into a more open-minded and accepting person.

Some of the debates raging today between farmers and vegans are no different to those two nine-year-old girls squabbling over their lunchboxes. Except, I'm pretty sure our debates were more mature.

Plant-based hysteria engulfed the food agenda around 2016/17. It was a laughable debate really, fuelled by a frenzied tabloid media which couldn't get enough of the binary 'Herbivore versus Carnivore' narrative. Animal rights hippies threaten white van man's bacon butties – yes! It seemed we only had two choices – either nail your colours to the vegan mast or fly the flag for flesh-eating. In the face of what felt like unrelenting abuse, farmers tried their best to fight back – ridiculing almond 'milk' and pointing out skinny vegans who looked 'a bit unhealthy'. But their arguments sounded hopelessly floppy and feeble against the terrifying killer question: 'HOW CAN YOU JUSTIFY KILLING AN ANIMAL THAT WANTS TO LIVE?'

I felt desperately sorry for livestock farmers around that time. The shock and panic I saw on the faces of many quiet, unassuming farming families, desperately grappling for a way to defend their very existence, their right to live. All this when more than 90% of households were still happily buying meat and dairy – confirming with their wallets every single day that we still need and want livestock farmers. It was a media-fuelled witch hunt – nothing less. It felt like the whole country had crowded around two kids in the playground shouting 'Scrap! Scrap! Scrap!', loving the entertainment while farmers got the shit kicked out of them.

But it has to be said, that emotionally blinkered dedication to the 'way of life' gets them into trouble sometimes. You cannot overstate a farmer's cultural compulsion to feed and sustain a nation. It is deeply ingrained – a sense of duty and pride handed down the generations. It is their reason for being. To say you don't want to eat their food – or worse, you don't want anyone else to eat their food – is like someone saying, 'your house is shit' or 'your parents are dicks'. And when it comes to the meat debate, there is a massive blind spot among livestock producers.

I get irritated by the livestock sector's institutionalised defensiveness and limpet-like adhesion to the status quo. I meet plenty of farmers with their fingers stuffed stubbornly in their ears, singing la-la-la in the blind hope that all vegans will get ill from B12 deficiency and come crawling back begging for steak.

One evening during my Nuffield Scholarship travels in 2016, nearing the height of vegan hysteria, I told a group of international farmer friends that I had chosen to have a meat-free day and voiced my opinion that everyone should have at least one meat- free meal, every week. A casual conversation around the dinner table, somewhere in Eastern Europe, very quickly turned into a debate, and ultimately into a row. A couple of the livestock farmers were particularly annoyed and pulled me up for saying 'everyone'; pointing out that it wasn't my place to tell other people what to eat. Fair point, but I refused to back down. Being in the minority and now firmly on the defensive, I fought my corner with increasing ferocity. Tempers rising, the argument grew ever more ridiculous. A Canadian cattle rancher bellowed: 'I want kids to eat beef for breakfast!' An Australian poultry farmer exclaimed: 'What about the body builders? They need meat every day!'

People got upset and some left the table. In hindsight, it probably had as much to do with jetlag and tiredness as our personal views on meat consumption. Nonetheless, it was a gruelling experience.

We all made up the following day of course – I hugged it out with the Canadian rancher in the hotel lobby and we all moved on. But it got me thinking about what had gone wrong. For me, it was a theoretical debate. How easy it was for me to proclaim: 'Everyone in the world should do this!' For them, it was personal – a direct threat to their livelihood and their identity. In hindsight, though I still don't agree kids should eat beef for breakfast and I'm pretty sure body builders can survive a day without meat, I understood their anger and annoyance.

But it is an inarguable fact that many of us in the West eat too much meat. In my social circle – so this is purely anecdotal – it tends to be the older generation and men who are most addicted to meat. My Mum and Dad eat it every day without fail. A meal without meat is a meal without heart in our house. I once cooked Dad a sweet potato and butternut squash chilli. He poked it around his plate for a bit before eyeing me suspiciously: 'What's this orange stuff?'

The other half of the butternut squash languished in their fridge for weeks. Sometimes Dad would encounter it, shoved at the back, while fishing around for a yoghurt. 'What's this bloody orange thing still doing here?' he'd grumble.

'It's Anna's butternut squash,' Mum would shout back. 'She left it for us, but I don't know what to do with it.'

But I, my two sisters and many of my (mostly female) friends, are increasingly following a flexitarian or plant-based diet, and progressive livestock farmers have already accepted this shift in consumer behaviour. In 2019 I can't hide my surprise when my good friend Rob Mercer, a major pig producer in Staffordshire, tells me he's selling off his hugely successful business to Cranswick, one of the biggest meat processors and food companies in the country. He casually drops it into conversation that he's also stopped eating meat five days a week. Coming from a 6ft 3in red-haired pig farmer, who's built like Tormund the Wildling from *Game of Thrones*, I'm not sure which shocks me most. When I ask, simply, 'why?' this is what he says:

'More people have, and are, moving to reduced meat diets, so we want to produce less meat but produce it in a way that's even more sustainable for the environment and better for the animals' welfare. Basically, we want to remain relevant to those customers who want to eat less meat, but meat they can absolutely believe in.' He's since bought himself a new farm in Shropshire and is pioneering an indoor/outdoor system for fattening pigs. He'll farm at lower stocking densities than before in large open-sided, straw-bedded barns. The pigs can choose to stay inside if it's cold, wet, or boiling hot, or wander out to the paddock if they fancy taking the air. Rob isn't afraid of change. In fact, he embraces it with a big old bear hug.

The fact is, irrespective of your views on whether it's right or wrong to eat animals, we simply do not need to consume meat at the levels we currently do, especially in the Fat West – those developed economies where meat has become such a cheap and widely-available commodity we can shove it into our gobs without a moment's thought for the living animal it once was, or the farmer who worked so hard to raise it. I've done it myself – dashing out of the office at lunchtime, grabbing a £2 ham and cheese sandwich off the shelf in a Tesco Metro and mindlessly shovelling it down without a moment's conscious enjoyment or appreciation. Surely a pig's life is worth more reverence than that? Meat is a gift – precious, hard-won sustenance our hunter/ gatherer ancestors would have given thanks for. I liken it to Daniel Day-Lewis in *The Last of the Mohicans* when he hunts and kills an elk and says a little thank you prayer. I'm getting much more Mohican about meat. Not that I stand in Tesco blessing the ham sandwiches, but I might choose veggie in a hurry and leave the meat as a treat, for when I have time to cook it properly and savour it.

I'm not sure how easy this would be if there weren't so many exciting meat-free options available these days. Our mainstream national menu has wholly transformed, even in my lifetime (and I'm not that old).

In school in the late 1990s, when I was doing my Duke of Edinburgh Gold Award, we were discussing what food we should take on our long-distance hikes. One of the girls in my group suggested 'Pestopasta' to which the others nodded enthusiastically. I was given the job of buying some. I had no idea what 'Pestopasta' was but didn't want to appear unsophisticated so popped it on the shopping list, unsure even of how to spell it.

Mum didn't have a clue either but in her ever-reassuring way she told me not to worry. 'I'm going to town tomorrow,' she said. 'I'll find some for you.'

Poor Mum walked aimlessly around Sainsburys in Oswestry not knowing where to start. She asked one of the assistants stacking shelves who didn't know either. When she eventually tracked down a jar of unappetising green gloop, she double-checked the label, shrugged, and popped it in her basket.

'Well, I found your pesto,' she said that evening, 'but I have no idea what you do with it.'

These days we've always got a jar of pesto in the cupboard, along with packets of red split lentils, bulgur wheat, spelt and quinoa, and cans of chickpeas, black beans, butter beans, and kidney beans – none of which I ate as a child.

Reducing meat consumption is not just an ideological decision, it is a natural transition made possible by the tasty and affordable plant-based alternatives that are now much more widely available. I was raised on a traditional British diet of meat and two veg – and grew healthy and strong on it – but I'm no less healthy and strong for switching some of the meat for beans, pulses, nuts, grains and 'that orange stuff'. The most important thing to me is knowing how my food – be it meaty or 'planty' – is produced, and what's in it.

In August 2020 I took our dog Lucy for a walk along the Harbourside in Bristol. There was an Extinction Rebellion protest underway, though lacking its usual vigour due to Coronavirus social distancing rules. I decided to grab some takeaway food and people watch, so wandered over to one of the trendy eateries on Wapping Wharf and perused a menu of wraps over a Covid-secure hatch.

'What's the vegan jerk chicken made of ?' I asked the girl on the other side.

'Satan' 'Excuse me?' 'Sei-tan'

'What's that?'

'I don't know. Gluten? Like pure gluten.' 'Gluten?'

'Yeah . . . so maybe . . . wheat?'

'Ok, so it's not a bean or a pulse, like soya or chickpea?' 'No, no I don't think so.'

'Um, I'll have the jerk chicken wrap please. At least I know what that is.'

Google made seitan sound just as unappetising by blandly confirming it was, indeed, a dense mass of wheat gluten. I gave it a second chance in the pub near our allotment a few weeks later by ordering a seitan burger. It came deep-fried in breadcrumbs and was truly, truly horrible. It tasted literally of nothing (I do not use the word 'literally' lightly, but it feels justified here). Ironically for a vegan meal it was like chewing leather. And cost me £10.

In October 2020, Alex and I had a night out – a rare and wondrous thing in between lockdowns. We creaked open our wardrobes, shook the dust off our going-out clothes and booked a table at a lovely restaurant in Bristol which prides itself on serving local and seasonal produce. Perusing the starter menu, my eyes were immediately drawn to the 'Heart of Palm with vegan black pudding'.

This was another new one on me.

'What is Heart of Palm?' I asked the young waiter.

Just like the girl behind the hatch in the wrap place, he looked very unsure.

'It's like a tube,' he ventured. 'It's a bit strange.'

The prospect of a strange tube on my plate, like some kind of edible tampon applicator, wasn't really selling it, so he went to ask the chef.

'It's the middle of a palm tree,' he announced triumphantly on his return, visibly satisfied with this newly acquired knowledge.

I decided to order it and, this time, I was not disappointed. The heart of palm was delicious but totally eclipsed by the vegan black pudding, which tasted just like the real thing. Again, I was full of questions – how can a plant taste like pig's blood?

As our friendly waiter cleared the plates, I asked him what was in it, and felt instantly bad for putting him on the spot again.

'I don't know . . . I'll check.'

Off he scurried to the kitchen, no doubt cursing his luck for getting lumbered with me on a busy Saturday night. It was a while before he came back:

'The chefs didn't know either – they had to check the packet. It's a beetroot and soya protein mix.'

I thanked him and dared to ask one last question: 'Do vegan customers ever ask you about the ingredients in this food?'

'No,' he smiled, as if the thought had only just struck him, 'they're happy so long as they see 've' on the menu. No questions asked.'

I took that as my cue to shut up and tuck into my main course – a beefburger which tasted a hell of a lot better than seitan.

For me, the vegan culinary experience is like a new shopping mall or a redeveloped city centre – shiny and exciting but still a bit soulless and sterile. Beyond ethics, vegan food is yet to tell its 'story' and find its place in our national feasts and celebrations, like a turkey at Christmas, a haggis on Burn's Night, mutton for Eid or a leg of new season lamb at Easter. That will change though, and I don't see why it's a threat or an insult to our farming industry, or worth getting our knickers in a twist about. Farmers will continue to produce everything I eat – whether it's beef or broccoli.

Far from being a negative, these well-meaning dietary choices at least show that we're thinking about food, and we're trying to buy what we believe in. Not everyone is so lucky. There are far more pressing issues in food and farming than whether you can milk an almond or not.

In September 2020, the Food Foundation published its Broken Plate report which examined what's working and what's failing in the UK food system. I learned with horror that diet is the biggest risk factor for disease in England. Considering I read this in lockdown at the height of a global pandemic when we couldn't leave our homes for fear of spreading a deadly virus, it came as quite a shock that a poor diet is more likely to make me seriously ill than Covid-19.

In short, it can kill you: either directly, by starving or over-eating yourself to death, or indirectly, by increasing your chances of developing heart disease, certain types of cancer or diabetes. The reason we don't all panic in the face of such a deadly threat and announce a national lockdown forcing us all to stay indoors eating salad, is because we can dodge this fate. We can avoid unhealthy food and buy healthy food instead. It is within our power not to die from a poor diet.

Except for millions of people, it isn't. Around the world, more than two billion people are as powerless against the epidemic of malnutrition as we all were in the darkest days of the pandemic.

A sad unifier of the urban and rural experience is that hunger and malnutrition do not discriminate based on your postcode. There are families everywhere either struggling to afford enough food or get access to nutritious food, or both, and it's a problem going from bad to worse: In 2020/21, 4.2 million people were living in food poverty. In little over a year, millions more had been dragged into food insecurity by the cost-of-living crisis. In September 2022, a YouGov survey found one in four households with children experienced food insecurity that month. That's almost 20 times as many hungry children as there are farmers. Shamefully, our levels of food poverty are the highest in Europe. Nearly a fifth of children under the age of 15 live with an adult who is moderately or severely food insecure.

These are shocking statistics but worryingly easy to ignore because the issue does not affect the majority. We know it's there, and feel bad about it, but the idea of going a whole day without eating, in our country – where food is still plentiful and affordable for most is for many difficult to grasp. I count myself in that.

We didn't have a lot growing up, but we never once went hungry. I have no idea what proper hunger feels like. From a tiny age I knew the difference between healthy food and 'rubbish'. I never worried about my next meal – it just kind of magically appeared. I understood why Mum put fruit on the table but kept the Milky Ways in a box I couldn't reach. These are the basic building blocks upon which your whole attitude towards eating is set for the rest of your life. As easy as breathing.

Most of us, luckily, share a broadly similar childhood experience which makes it difficult to fathom a food insecure existence, and maybe that's why we've been so crap at tackling the problem. We can sympathise but we can't really understand.

I didn't get it until I interviewed Karen Washington outside a tomato polytunnel in New York.

Karen is a farmer and food justice campaigner who has determinedly dragged fresh fruit and vegetables into the North-West Bronx, one of the poorest neighbourhoods in New York City, by setting up the hugely successful weekly farmers' market La Familia Verde. It wasn't the community who made this difficult but doubting outsiders, including farmers and growers, who simply didn't believe less affluent people would buy their produce. Bit by bit, Karen is smashing the myth that low-income neighbourhoods can't afford fruit and veg:

'You put the group of poor people together and they spend more money per capita than any other group, so that's number one . . .'

She lifts her thumb to emphasise the point, swiftly followed by her index finger. Myth Number Two: that these communities would rather live in a food desert of fried chicken.

'No, no, no!' She looks furious. 'When I first heard that term I was mystfied. Living in a low-income neighbourhood no one uses that term. Food desert? What does that even mean?'

It's a warm October day on Rise & Root Farm on the edge of Chester, a small rural town about 50 miles from Manhattan. Sprinklers are spitting away in the greenhouse, drenching neat rows of leafy greens with rhythmic bursts of spray. It's a cathartic sound after the noise of New York and I close my eyes for the briefest moment, enjoying the warm humidity.

Karen is a born and bred New Yorker with a former life as a physiotherapist. She's been farming this five-acre patch of land on the black dirt soils of Orange County since 2015, growing organic vegetables, potted plants and edible flowers which get loaded up in a van and trucked over the George Washington Bridge into NYC, where her farming career began.

Conversations over the raised beds in the community gardens of New York not only taught Karen how to grow plants and harvest crops: it was where she found her voice. She learned more about the social problems in her community – people struggling to pay rent, not being able to afford heating and hot water, police brutality and drug abuse – and saw how it was all connected to food:

'Food intersects so many parameters.'

She ticks them off: 'Our health with diet-related diseases. When it came to housing issues, a lot of people were not feeding themselves because they had to pay rent. Then there's the relationship between food and education – kids were going to school hungry. There was overcrowding in schools and sometimes kids were getting lunch at 10 or 11 o'clock and then coming home hungry. I saw the relationship food had to all those social issues.'

As Karen lays it all out there, I start to understand the deadly tentacles of food insecurity; how it wraps itself around everything, suffocating a child's chances of succeeding at school, and then in work and, ultimately, in life.

I grasp for an analogy – how can I explain what this is like?

'I equate it to modern-day slavery,' says Karen, who is African American. 'That's how I look at it. Especially in my neighbourhood. When we came to this country enslaved, we were given the scraps, the food that nobody wanted. Now here we are in the twenty-first century and we're still given the food that nobody wants. It's like we're repeating history again. We're enslaved to the processed food, the junk food and the fast food.'

I don't know what to say. There's nothing I can say – white privilege is plastered all over my face. I let her words sink in.

I'm not sure how to ask my next question. I want to know Karen's thoughts on food poverty in rural areas; white America basically – as far from the Bronx as you can possibly get. Immediately I catch a spark of interest, a subtle double take. Karen has been interviewed a thousand times before about setting up farmers' markets in the city. She's got some well-rehearsed answers to the same old questions and may well punch the next journalist who asks about 'food deserts'. I guess I'm asking her something different.

'You know,' she says thoughtfully, 'we talk about "the poor" in the food system and people think about urban areas where mostly Black and Latino people live but, believe it or not, the majority of low-income people are actually white that live in rural communities. Poverty knows no colour. When we're talking about poverty, we need to make sure we're not lumping it all in urban areas. A lot of it has to do with people's perception of what they think poor looks like.' Food poverty in urban areas is undoubtedly more intensified, clustered in smaller areas and affects larger groups of people, but it also has the advantage of being more visible. And when we can see a problem, amazing people like Karen, and hopefully governments,

will have a go at solving it.

But driving across Illinois, Iowa, and South Dakota, it's more about what I can't see. What's missing. It's a lot easier to buy a salad in New York or Chicago than on the road in deeply rural areas of the Midwest. The breakfast buffets in budget hotels are, without exception, a selection of sugary cereals, highly processed juice, cheap bread, waffles, pastries, plasticky cheese and processed meat. On the rare occasions I find an apple it looks waxy and suspiciously shiny.

I've spent weeks on the road in the Midwest, on several different research trips, and no word of a lie – I can feel the lack of nutrition. My mouth tastes furry, my digestion is off, I'm bloated. I don't feel like me.

One morning, in a quiet roadside hotel just off Route 80, about halfway between Iowa City and Des Moines, I peruse the breakfast menu. There's a choice of eight breakfasts variously based around eggs, pork chop, steak, fries, cheese, buttermilk biscuits, vanilla egg batter, sausage gravy and a 'monster of a cinnamon roll'.

I scan the menu. No fresh fruit, no yoghurt, no muesli. My heart sinks. I genuinely feel a sensation of panic – knowing I need a nutritious breakfast, feeling my body flag on a solid diet of meat subs, fries, burritos, tacos, burgers, chips, candy and all the other processed rubbish I've been filling up on at gas stations.

I go with the best option available when the waitress comes to take my order:

'I'll have the oatmeal please.' 'Off the kids menu?'

'Yes please. I just want something light.'

'It's instant. I have maple and brown sugar, sugar and cinnamon or cinnamon sugar and spice.'

'Do you have plain oatmeal?' 'No.'

'Um. I'll have the least sugary one please.'

Each time I return home to the UK from a road trip in the US, I feel enormously thankful for the copious amounts of readily available fresh food. I love our food. It never crossed my mind that the same problems exist here.

Barbara Bray MBE is a registered nutritionist and food safety consultant and has worked in the agri-food sector for more than two decades. Her focus on fresh produce procurement means that she spends a lot of time in the east of England, big veg country, and lived in Lincolnshire for a while. It amazed her how difficult it was to buy fresh food in such a horticultural hotspot:

'I'd be standing outside my own front door and could see a carrot field to the left, broccoli field to the right and lettuce behind me, but I couldn't buy any of it very easily within walking distance or even a short drive away. That's one thing that really surprised me in the 20-odd years I've been doing my job. When I was on the road, I'd buy food from McDonald's, Subway, or the Shell petrol station. In the extremely rural areas, there wasn't even a fast-food place. You could maybe get some salad bits in the local village Co-op but the range was a lot smaller, and expensive. I found it's quite hard for a rural population to access healthy food because the shops they have aren't selling that kind of stuff.'

I'm reminded of Karen's words in New York: 'We're enslaved to the junk food.'

So too, it seems, are people in Blackpool, Burnley, Hull, Stoke-on-Trent, Great Yarmouth – towns, cities, villages up and down the length of our little islands. This isn't a far-away problem for 'those fat Americans' – we're in a terrible situation of our own making. And it only got worse during the pandemic.

Sally Mercer is married to Rob, the pig farmer and wildling lookalike.

'He also looks like the ginger Scotsman in *Braveheart*,' she adds helpfully.

Together they set up a charity called Farm Fresh Revolution, inspired by Rob's Nuffield Farming Scholarship in 2016 (that's how we first met). He was looking for ways that farmers, specifically, could help tackle food poverty. He travelled around the US, Brazil, and Europe meeting farmers who were using their knowledge, networks, and contacts to do incredible things:

'We have automatic trust thanks to that unique relationship between farmers and community,' says Sally: 'When we arrive at the schools, they say: 'The farmers are here! It's the farmer van!'

Rob and Sally's Farm Fresh Revolution is aimed at helping families with young children cook and eat healthier food in their home county of Staffordshire. They supply and deliver fresh fruit, vegetables, eggs, and meat to disadvantaged communities in Tamworth, Burton-on-Trent, and Stoke-on-Trent, working with seven primary schools, two community centres and a soup kitchen. Pre-Covid, they would set up fortnightly market stalls, run by two to three volunteers, outside the schools on a Friday afternoon just before pick-up time. Parents could come along and help themselves to a bag of groceries: onions, potatoes, carrots, green vegetables, salad and peppers in the summertime, apples, oranges, bananas, sometimes strawberries, free range eggs and some beef mince, sausages, or chicken drumsticks. Farm Fresh didn't charge a penny, but there was a donation box if people felt able to contribute.

'The kids would all come and flock around the table and grab a carrot or a satsuma or an apple, with a lolly in one hand but actually preferring to eat the fruit over the lolly. It just shows that if you can put the right things in front of a child, they will take that opportunity. If it's not there, they can't take it,' says Sally.

In lockdown, Farm Fresh switched to weekly deliveries, sending out 300 pre-packed bags a week. Rob and Sally watched in amazement as the pandemic brought them a step closer to their ultimate goal – inspiring healthier relationships with food:

'What we found over Covid is that a third more people cooked at home. Now, that's not a surprise for most of the population but, for the people we work with, a lot of them don't generally cook that much. They don't feel they know how to, or don't have the facilities or the time, or, in their eyes, it's not affordable to cook fresh food.' 'We give people recipes and occasionally the community centre will cook up the veg so people can taste it. Sometimes we throw in a curveball, like a butternut squash. One woman said, "No one's ever going to eat that," but we persuaded her to give it a go at home and now it's her son's favourite vegetable. Lots of parents tell us their kids used to be really fussy eaters but now eat all the vegetables we give them.'

Sally is visibly excited, and it lifts my spirits to hear there are positive trends out there, even among our most food insecure communities. Farm Fresh have given away nearly 27,000 bags of food which equates to more than 300,000 homecooked, healthy meals in just one county.

It brings hope – something to work with and build on, especially in the face of so many depressing statistics.

In April 2020 I directed a film for *Countryfile* (remotely from my kitchen) about the crisis in dairy farming, triggered by the closure of the hospitality sector. The loss of all those lattes sold in Costa, Starbucks and McDonalds led to a collapse in prices and an oversupply of milk. Thousands of litres were poured down the drain. Appalled, I started cycling out to Old Green Farm Dairy on the edge of Bristol, a cool bag stuffed in my pannier bags to buy milk fresh from their vending machine, paying £1.50 a litre instead of the 23p some farmers were being paid in the April 2020 lockdown. My first taste of raw milk, standing next to my bike on a sunny spring day, had an extraordinarily visceral effect on me, unlocking some deep, long-lost memory. I was transported – somewhere. I told Mum about it on the phone, to which she replied: 'Well, you were reared on unpasteurised milk until you were seven – Dad used to bring home a little churn every day when he milked cows at The Bryn.'

That was a long time ago – before Dad built his own business, when he was still working as a herdsman on a local dairy farm. I find it extraordinary that my tastebuds can remember something long forgotten in my conscious memory. All I had to do was unlock it.

CHAPTER SEVEN

ENVIRONMENT

IT'S A HOT JUNE DAY IN A Romanian hay meadow. I'm watching a Transylvanian count wading through thigh-high grasses peppered with ox-eye daisies, yellow rattle and sainfoin, hovering his mobile phone over the flowers playing a recording of a corncrake. 'Sometimes they answer back,' he says, and tips me off that you can simulate their rasping call by running a pen along a comb.

I've never heard a corncrake before, not even on a mobile phone. These secretive little birds, barely bigger than a blackbird, are related to moorhens and coots, but unlike their wet-footed cousins they tuck themselves away in tall vegetation, nesting in hayfields and meadows.

I'm sure Grandad Bill knew the call of the corncrake – they were widespread in the UK when he was a lad but declined catastrophically during the twentieth century when farmers switched from hand scything to mechanical mowing and cut their grass crops earlier in the year. Baby corncrakes, hunkered down in the grass, didn't stand a chance, but before you picture some maniacal 'Here's Johnny' moment on a Massey Ferguson, I'm not suggesting farmers like Grandad took any pleasure in this or even knew what was happening.

But it happened nonetheless, and the corncrake is now a red list species – a bird of high conservation concern – and your best chance of seeing one is on the western and northern islands of Scotland.

Yet here in this meadow bordering the forested slopes of the Carpathian Mountains, corncrakes abound; though I'd struggle to hear one over the thrumming cacophony of insect life, as chirruping crickets sing their hearts out in the sunshine.

Count Tibor Kálnoky's family has lived here for 800 years – not counting a 60-year hiatus when they were exiled under fascism and communism. In 1989, the aristocrat was able to reclaim some of the ancestral lands in Hungarian-speaking Transylvania and returned from exile in Paris to live here permanently in the 1990s. He's a smooth-looking guy, bearing a vague resemblance to Harrison Ford, and very jovial, though I've been warned not to make any cracks about Dracula. This is difficult when the Count excitedly introduces us to his rare breed flock of black Transylvanian bare-necked chickens, the gawkiest-looking birds I've ever seen with their long, naked necks protruding from black feathers like bright red periscopes, ripe for biting.

The Count's enthusiasm for the native birds and animals of Transylvania – be it a weird chicken or a shy corncrake – is infectious, and I find myself willing the birds to respond to his conscientious efforts with the mobile phone. He gives up and re-joins our team at the edge of the hayfield.

It is June 2015. A bumpy three-hour journey, traversing canyon-like potholes most of the way from Tirgu Mures airport, has brought us to the tiny village of Zalánpatak. I'm with Charlotte Smith, presenter of Radio 4's *Farming Today*, and Dimitri Houtart, the BBC's rural affairs editor, producing a radio programme about biodiversity with Count Kálnoky and his friend, His Majesty the King, then Prince of Wales.

King Charles first visited Transylvania in 1998 and was immediately captivated by the region's 'timelessness'. He said it reminded him of stories he read as a child – where bears and wolves roam the forests, mountain pastures tinkle with cowbells and clouds of silver studded blue butterflies dance to the swoosh of a farmer's scythe. For a man who has dedicated 50 years trying to get us to see the value of nature – here it was in all its priceless glory.

In the shade of a rustic open-sided barn, I nervously grapple with the heir to the throne's lapel, attaching a radio mic to record his interview with Charlotte Smith for this special edition of Radio 4's *On Your Farm*.

They talk about the immense pressures on our environment and the King's vision for a new form of accounting, which puts an economic value on our 'natural capital'. Think of it as a bank account for biodiversity – if you make withdrawals or deplete the balance in any way, you'll get hit with some big charges. If you leave it to rest and grow, humans can live off the interest in harmony with nature.

'Maintaining nature's economy,' he says, 'is absolutely central to maintaining our own, human economy.'

The King has dedicated half a century to environmental campaigning, driven by a desire to protect the biodiversity he and every other baby boomer inherited for his grandchildren's generation and beyond. At that point in the interview, he unfolds a piece of paper in his lap and asks if we would mind if he quoted from 'a marvellous piece of writing' by the British travel writer Robert Byron. He proceeds to read 'All These I Learnt', written sometime in the early twentieth century. The poem lists nearly 100 plants, insects, birds, and mammals that Byron knew as a child from 'butterflies that suck the brambles' to 'orchids, mauve-winged bees and claret-coloured flies climbing up from mottled leaves.'

The poem finishes with a tribute to nature and a promise to the future: 'All these I learnt when I was a child, and each recalls a place or occasion that might otherwise be lost ... They gave me a first content with the universe. Town-dwellers lack this intimate content, but my son shall have it!'

The King takes off his glasses, folds away the paper and says: 'All those things he wrote about, so many of them have gone in Britain. But they're here.'

I look out to the Transylvanian meadow, an antiquated pastoral scene that feels viscerally familiar. I feel like I've been here before – but that's impossible. It must be an echo of my humanity, my species. Because I have never seen anything like it in my lifetime.

There, on a summer's day in Romania listening to the future king read poetry, was the first time I felt what I now know is called 'eco grief'. My longing for something lost.

Since the 1930s we have lost 97% of wildflower meadows in the UK. Most were dug up, reseeded, and fertilised as part of a huge food production drive during and after World War Two. One ancient meadow after another disappeared under the plough in a matter of hours, as farmers answered the Government's call to grow more grain – replacing messy meadows of riotous colour with neat, golden rows of wheat and barley. And they ploughed with zeal, these farming families who had struggled to feed themselves, let alone the rest of the country, throughout the Great Agricultural Depression which had dragged on from the 1870s. The slump had lasted so long – well over half a century – that Britain's farmers, many of them poor tenants like my ancestors, just got used to their irrelevance. They despondently plugged away on their little farms as dazzlingly cheap and plentiful grain poured into the docks from the newly planted virgin prairies of the American Midwest. It kept the ballooning urban population fed – an industrial food system for an industrial workforce – and that kept the government happy. Job done.

A brief reprieve for the neglected farmers came in 1914 with the outbreak of World War One when millions of meals ended up at the bottom of the Atlantic, as German U-boats sank every merchant ship they could. Fearing food shortages, British politicians suddenly remembered they had farmers who could grow things, which came in very handy for four years of war, but they were predictably dropped like hot potatoes in peacetime, when grain shipments resumed.

With a second war came a second chance. This time, united under a newly emboldened lobbying force in the form of the National Farmers' Union, landowners and tenants alike were ready to do what was needed. They ploughed for pride and profit and were ignored no longer. Post-1945, with the country still hungry and rationing remaining in place for another nine years, far from being dropped, farmers became the focus of new science and engineering innovation – agrochemicals to protect crops and boost yields and heavier tractors for faster work across bigger areas. Farmers were restored to a position of prominence, influence and, for some, even wealth. Government subsidies incentivised production and big business clamoured to work with this new and growing market, developing more and more chemicals, pharmaceuticals, machines, and gadgets. They were joining the 'Boys of Big Ag' in North America – the food producing giant whose exports had pummelled Britain into agricultural insignificance for decades. Finally, British farmers had a seat at the table. They felt relevant again.

The rich biodiversity that quietly sat in the background was suddenly catapulted into a new era of chemical and mechanical progress. Wildlife had flourished during British agriculture's long sleep. Overgrown hedgerows, boggy fields of tufty grass, dilapidated buildings and general weedy, brambly untidiness was a big ol' biodiversity party. Not that anyone called it 'biodiversity' then – instead it was defined by the birds, insects, animals, flowers, trees, and hedgerows that every farmer and labourer knew by name or, at least, had their own names for. As children, Dad would tell us that the trees are our streets. Just as urban dwellers navigate towns and cities by road names, we should read and understand nature's signposts. I've never been particularly good at it though and still resort to random guesses whenever he skewers a fallen autumn leaf on the end of a stick and holds it up for inspection: 'What's this Anna? Well, come on, you know what this is! Beech!'

As lovely as the meadows and hedgerows were, they hadn't put money in the bank through those decades of hardship and struggle. Their economic value was hidden. Landowners just saw outdated, backward farms being left behind in a changing world. There wasn't room for sentimentality. The meadows were ploughed, the hedges ripped out. Farms and fields got bigger as biodiversity ever so gradually dwindled.

Only an insane person would set out to 'kill nature' but, equally, back then, very few farmers and land managers considered protecting it. Why should they? Nature was an omnipresent, insuppressible force. Like God Himself. It gave rain and sunshine and made the crops and animals grow, but it also brought floods and drought and deadly diseases which made their crops and animals die. Nature could not be dominated or extinguished – despite their best efforts at times to catch an industrious mole ruining a hayfield or a wily fox killing their chickens. No, nature can look after herself – we have a job to do.

I've spoken to many older farmers who remember those days of agricultural transformation in the second half of the twentieth century. An East Anglian arable farmer in his seventies told me wildlife was 'adversarial' and remembers when skylarks were a 'terrible nuisance' pecking at newly sown crops and flying about the tractors. He vigorously embraced the new era, draining and planting wetlands, even hiring an aeroplane during the 1970s to spray the farm with chemicals, which killed all the plants in his wife's garden.

You can track this fascinating period of intense agricultural change simply by talking to different generations of the same farming family. To understand where farming has been, and where it's going in the future, I recommend doing it sometime – find a family farm and walk through its timeline with them.

Every field has a life story. Like Spratts in Essex.

It's early March 2020 and George Young sinks into the claggy blue London clay. A long winter of storms and incessant rain has turned the wheat fields into mud flats. Yanking his welly from the sucking gloop, George knows it'll be weeks before they can get this field planted. Spratts is a mess.

The field got its name – Spratts – when George's father and grandfather pulled up the wych elm hedgerows in the early 1970s, turning what used to be four fields into one much more productive, 48-acre field. It allowed new and larger agricultural machinery to move easily across the land. The tired old hedgerows, ravaged by Dutch elm disease, surrendered to agricultural advancement without resistance. At the touch of the chainsaw, long lines of hollow deadwood shattered to the ground. Piled high and tinder dry, it burned fast and bright – the embers of the old making way for the new.

The newly amalgamated Spratts was fertile and bountiful. In the early years, potatoes, peas, wheat, barley, oilseed rape and hay for cattle were grown in it. Gradually, over time, the rotations reduced. Pea picking, which used to be carried by local housewives at harvest time, disappeared from the farm in 1985. The last potato was harvested in 2005 and dwindling numbers of cattle were pushed out to graze on the nearby marshes. Spratts increasingly became devoted to specialist crops of winter wheat and oilseed rape, with the occasional break for spring peas (no longer picked fresh but left to dry in the field before being combined). In an intensive six-year rotation, despite heavy clay soils, it would faithfully produce four good crops of wheat. Spratts is a workhorse of a field.

It hadn't always been destined for agriculture. Had George's great-grandfather not come along in the early 1950s, Spratts and the rest of the farm would have become a dumping ground. The land's proximity to London, in spitting distance of the River Thames and what is now the London Gateway Port, made it the perfect location for industrial waste disposal. It was zoned for this purpose in the 1930s but problems with contracts and pricing meant the owners had to put the land up for sale instead.

Farmers through and through, the Youngs are descended from Scottish dairy farming stock. In 1895, George's great-great-grandfather, James, put his cows on a train from East Ayrshire to the south of England to supply milk into the lucrative London market. Fast forward half a century and the family were ready to move into cereal and vegetable production.

In 1954, James' son, Andrew Gemmill Young (George's great-grandfather, who I'll call Andrew 1) was looking to buy a farm with his two sons Jim, a dairy farmer, and Andrew 2 – a soldier looking to escape Army life after fighting in Burma in World War Two. Two hundred and thirty acres of Essex marshland, a house and three farmworker's cottages would do perfectly. The Youngs purchased Curtis Farm for £10,000.

Throughout the 1950s and 1960s they drained the marshes and grew fresh produce to sell in London's markets – runner beans, marrows, picking peas and, most profitably of all, spring greens.

In 1968, the next generation entered the business – confusingly, yet another son called Andrew. Andrew 3, George's father.

Now 69, he remembers those early days on the farm with fondness: 'It was exciting,' he says. 'You could grow a crop and send it up to London, to the market at Spitalfields, and if there was a shortage of that particular thing, you could make a killing. I remember a time when a crop of spring greens was worth more than the value of the land.'

But farming changed rapidly in the second half of the twentieth century. As the sun set on Andrew 2's time, a veteran who'd known the food scarcities of wartime, the backbreaking reality of pre-tractor agriculture and the simplicity of selling fresh produce direct to Londoners, the sun was rising on Andrew 3's farming career. It coincided with the so-called 'Green Revolution' and the development of new, high-yielding cereal varieties, labour-saving mechanisation and miracle-working chemicals that obliterated pests and weeds with one easy application. It was transformational. Supermarkets would soon revolutionise food supply chains allowing the Youngs to sell commodities from the farmgate instead of schlepping up to London and sitting behind a market stall.

Yields rocketed and wheat, by far the most valuable crop, started to dominate the Youngs' farming enterprise. When the UK joined the European Economic Community in 1973, wheat wasn't just profitable – it was subsidised. Hunger was still well within living memory and European leaders wanted absolute assurance that the population could be fed. They were willing to pay handsomely for peace of mind under the Common Agricultural Policy and ambitious young farmers like Andrew 3 were more than happy to step up production.

'1984 was a particularly good year,' he remembers. 'We had huge crops of wheat, guaranteed prices and you couldn't fail to make money. And so it went on, until we ended up with butter mountains and grain mountains. It was utter madness but as farmers we didn't see that. We were being encouraged to do it and we were making a good living. We didn't question it.'

Spratts was there through it all – yielding crop after crop and responding well to the fertilisers and chemicals. Andrew 3 ploughed, planted, and harvested Spratts year in, year out and, for 30 years, the field performed beautifully. It was well into the twenty-first century before he noticed a change:

'When you're out there every day you see the soil and you can see the fields with good soil, and I thought Spratts was a good field. But there was one year, in about 2010, when we ploughed it and it looked dreadful. The soil didn't move like it used to. It didn't have any life left in it.'

The soils were exhausted, yields had plateaued, and the grain market was volatile, with prices collapsing more often than they were peaking. Curtis Farm, only a decade into the twenty-first century, barely resembled the land Andrew 3 had known in the middle of the twentieth century:

'I enjoyed my farming, but looking at things now I'm not sure we did the right things. A lot of what my generation have done in terms of food production was overkill.'

But, even with hindsight, he doubts he would have done things differently. A product of his time, a pupil of productivism – the peer pressure was to move with the times:

'The technology had moved on. You had to use fertilisers and if you didn't you were going to be at a disadvantage. But when you start using chemicals, you have to keep using more.'

The razzle-dazzle of miracle-working agrochemicals had outshone the miracle of nature which had, over millennia, quietly created all that life in the soil under Spratts, and which Andrew 3 found to be so depleted in 2010. Farmers, who had done everything by the book, followed all their training from agricultural college and listened carefully to the wisdom of agronomists, were as confused as everyone else. What had gone wrong?

Farming and nature have lived side by side since the dawn of the agricultural age 10,000 years ago. They have evolved to co-exist, sometimes in a beautiful partnership, but by the 1960s and 70s cracks in their relationship were starting to show, like an old married couple drifting apart. It was barely discernible at first. Out in the fields, farmers were still working alongside the birds, insects and animals, the flowers, trees and hedgerows, every day. OK, maybe not as many as there used to be – but they seemed to be doing just fine. They couldn't see, or perhaps chose not to see, that the delicate power balance was tipping dangerously. Old Macdonald had become a busy and successful career man while the wife he took for granted, Mother Nature, found it harder and harder to thrive.

As with so many broken relationships, outsiders could see it – and they started making noise about it. The modern environmental movement gathered momentum in the 1960s, fuelled by the publication of Rachel Carson's ground-breaking book *Silent Spring*, which pointed the finger at agricultural chemicals for the decline in wildlife and bird populations in the United States. As far back as 1962, Carson was warning of chemically-resistant insects and weeds. Her words have come to pass time and time again: 'The chemical war is never won, and all life is caught in its violent crossfire.'

The politically influential farming unions and increasingly well-funded conservation organisations locked horns, like love rivals.

'I know what's best for my wife!' hollered the farmers.

'You don't appreciate her!' cried the conversationists.

Frustrated at the inaction from government and the lack of engagement from farmers, the environmentalists went directly to the public and soon figured out that an emotional crisis narrative caught the attention of the media. For decades, farm groups and green groups have waged war on two fronts – in private, through traditional political lobbying, and in public, through the media. In 2017, as part of my Nuffield Farming Scholarship research, I visited the Brussels office of Friends of the Earth Europe and met with biodiversity campaigner Robbie Blake. He told me the farming unions are hard to beat in the corridors of power:

'There is a saying in Brussels that the farming lobby is the most powerful in the city,' he said with unmistakably raised eyebrows.

If you can't get to the politicians, take your fight to the people. If the media is a weapon, environmental groups wield it like public relations Samurais:

'There's very much a sense that the bad news story or the big threat is always the thing that's effective at getting a response in the media,'

Robbie told me. 'What would induce someone to write a letter, or sign a petition if it's not pressing and urgent? And we do that in good conscience – these are issues that need to be addressed and need to be part of public debate.'

And farmers take it very, very personally.

Even now, more than 75 years after VE Day, there remains a tangible sense of betrayal. After everything farmers had done to feed the nation through two wars and rationing? Their sweat and labour, eye-watering levels of investment and debt, all to produce affordable food for the consumers who'd ignored them for centuries and were now stabbing them in the back? A deep bitterness runs through the bones of farmers all over the world.

Adrienne Attorp is a Canadian environmental activist living in Newcastle. She moved to the UK in 2011 from a small town in British Columbia and spent a lot of time working on urban farms in London. She is now a Sociology and Social Policy PhD researcher, heavily involved with several citizens groups and the student activist movement. She's also spent a lot of time interviewing farmers in Northern Ireland and the Republic of Ireland for her research into agricultural pollution, meaning she's asked some awkward questions about the environmental impact of their farming practices:

'They see themselves as scapegoated for all the environmental problems,' she says, 'I remember sitting with one farmer and he looked up at the sky and said, "You see that plane up there? How many carbon emissions are coming from that? Fine, we've got 150 cows, but I also have trees and hedgerows and whatever."'

Adrienne pauses and ponders for a moment, weighing up whether the farmer had a point or not.

'I think with farmers it is a bit different,' she decides, 'because they are making something we need. Flying is often a luxury really, so is owning a car, whereas we need food to survive. That's where farmers have a slightly different leg to stand on and maybe that's why they're indignant. Like: What do you expect? We are keeping you alive, we are allowing you to survive!'

There's been competitive rivalry between the 'Fergs' and the 'Greenies' for as long as I can remember. Townies, tree huggers, hippies, bunny lovers, veggies – we'd bandy these terms around as kids and it never occurred to me it was a symptom of the divide. It was just how people spoke in our community. Often, it was meant affectionately. I'd listen to the old farmers chatting in the market: 'Aye, he's a nice 'owd lad; bit of a townie type but nice enough.' Sometimes it was an expression of frustration. I remember Dad reaching for the remote whenever a Linda McCartney ready meal advert came on: 'What are these bloody veggies on about now?' he'd grump, before switching the channel.

And, of course, there were all the usual nicknames for us farm kids at school: 'sheep shaggers', 'straw-chewers', 'Fergs', 'inbreds' – 'hurdle bunter' is a new one I heard recently. At university people asked if Dad chased people off his land with his gun. My best mates still call him 'The Cowboy' picturing Dad, a lifelong John Wayne fan, going round his sheep on horseback. But the teasing always made me smile. I've never felt a moment of real malice or aggression – even the old classic about Welsh farmers sticking their favourite sheep's back legs in their wellies to hold them still.

This has changed in recent years. Both sides are consumed with rage, even hatred. Fergs have become 'factory farmers'; the Greenies 'militant extremists'.

On the face of it, this is not an urban/rural issue – it's more about the hardening of a decades-long conflict between agriculture and conservation, which transcends geography and demographics. But urban/rural cultural tensions are there, simmering beneath the surface of the environmental debate. Agriculture is still, overwhelmingly, a rural, land-based industry embedded in a politically and socially conservative culture and, along with its large migrant workforce, employs a lot of people who have been born and bred in the countryside. The lifestyle choices to help tackle the climate emergency that are so readily available to us in an urban environment, such as switching to an electric car or cycling everywhere, are simply not an option for these isolated, often low-income rural communities. Employment opportunities are thin on the ground – to the point that going down the pit in a new Cumbrian coal mine could even be seen as an attractive option to some.

Meanwhile, the new wave of environmentalism, in the form of climate action groups like Extinction Rebellion and the political activist wing of the vegan movement, has mobilised the urban youth. Though the founding mothers and fathers are a diverse mix of urban and rural folk, their ideals are embedded in left-wing, socialist, anarchist culture – which is most at home in our cities. As I write these words, as if on cue, Johnny Rotten's on the radio: 'I wanna be anarchy in the city.'

Yeah, Extinction Rebellion is a bit like the Sex Pistols. Or so I thought.

I join an Extinction Rebellion Zoom call one evening during lockdown. I'm in a very recruitable frame of mind, so my interest is genuine.

The 20 to 30 smiling faces on my screen are the friendliest, politest rebels you could ever wish to meet. There's a warm ripple of 'Welcome Annas' as the meeting gets underway. It feels like a parish council. A lovely thespian-looking lady reclining on a sofa with some embroidery on her lap updates everyone on her working group's media and messaging work. Parents pop up with their kids in the background. I hear about the XR cricket club, and a disembodied voice asks: 'Is our samba band rep here?'

Now and again, several people spontaneously do jazz hands, I later learn this is instead of clapping, which can interrupt people or ruin their flow.

Even the civil disobedience bit is like planning a church fete:

'If you're up for some mid-spiciness we'll be occupying a financial institution on ... [I'll keep the dates confidential].'

'Oh, I can't sorry, I've already booked my holiday then.'

As the discussion gets spicier (XR rank their activities by level of 'spice'), it dawns on me that, if I wasn't a rebel in my teens, I'm certainly not a rebel at nearly 40. I'm not up for sitting on roads and runways. I wish I was, but I'm just not. I volunteer to put some stickers on lampposts, but no one takes me up on it.

One of my closest and dearest friends, Laura, puts the 'active' into activist. She's a member of Extinction Rebellion and has set up two climate action groups in South Gloucestershire. We marched together through Bristol city centre in the pouring rain with Greta Thunberg in February 2020. Laura, soaked to the skin, wet hair plastered across her face, looked so alive – shouting: 'What do we want? CLIMATE ACTION! When do we want it? NOW!'

She has dedicated herself to the cause, commuting to work on an electric bike, buying a hybrid family car and pursuing a plant-based diet as much as possible, which isn't easy with two young children. Climate justice has become part of her identity. She is emotional, angry, motivated, and determined to drive change:

'I'm severely worried about what the future holds in a world of climate chaos and I'm worried about my children's future,' says Laura over Facebook Messenger, which dilutes none of her fieriness. 'Working in academia, my job is to communicate climate change and air pollution – so the reason I started to take so much climate action is because I was feeling hypocritical by communicating it to other people but not taking action myself.'

'Here's another good slogan…' she adds. 'People gonna rise like water, gonna turn this system round, hear the voice of my great-granddaughter, saying: Climate Justice Now!'

I admire Laura's climate activism – but something unspoken prevents me from jumping feet first into her world. I feel deeply, squirmingly self-conscious and conspicuous. Like there's a giant flashing neon sign above my head proclaiming: 'BEEF FARMER'S DAUGHTER!'

At Laura's fortieth birthday party, all the sausages were meat-free. It seemed so overt. I felt like I wreaked of carnivorism: 'You don't belong here. This is a buffet for people who care about the planet.'

Of course, this was all in my head and nothing to do with my friends at all, who put on a lovely spread. But it got me wondering – am I just a bit weird and awkward or do other people feel like this too? Is this a symptom of the divide?

I talk to Ruth Grice about it, a dairy farmer who also works for Nottinghamshire Wildlife Trust. She has a foot in both camps too, running the family farm in rural Leicestershire and working for a conservation organisation in the centre of Nottingham.

'When I'm in the office it could not be any more different to being on the farm,' she says. 'It's in quite a deprived area of the city and I see people that I would never normally meet. The same applies to my colleagues. A lot of them live in the city and have very strong views about animal welfare and intensive farming. My desk in the office is next to the breakout area where we have lunch and coffees. It's not unusual to overhear conversations and it is upsetting when you hear myths about how people operate on farms. I sit there thinking, Oh my God, are they thinking I'm like that on my farm?'

But Ruth stays quiet. She doesn't say a word. Why not?

'They are my colleagues and my friends,' she says, 'and if any of them wanted a walk around the farm I would gladly show them but, in that particular situation, it just doesn't feel like the right thing to do, to just wade in and say, "Well, we're not like that." I don't want to come across as if I'm trying to change their minds because I respect their views. I don't like it when people are fed mistruth and believe that it's real but, equally, I wouldn't want to come across as defensive.'

Ruth has nailed it. That's exactly how I felt at Laura's plant-based birthday buffet. I am hyperaware of my personal attachment to animal agriculture and worry that the slightest comment could come off as 'anti-environment' – which isn't who I am. But equally, there is an instinct to defend, hidden beneath my silence.

Ruth feels the same: 'It's very emotional and close to home. I can't help but take it as a personal attack sometimes.'

So, to avoid any discomfort or confrontation, we stay quiet. Safer to say nothing. Silenced by our own self-consciousness.

This is hardly a surprise coming from two farmer's daughters, but I was shocked to learn even some environmental activists have felt out in the cold on this one. Sociologist Adrienne Attorp is involved with several citizens groups and the student environmental activist movement in Newcastle. She's also a meat eater. But doesn't really talk about that bit.

'I try not to engage in what people should and shouldn't eat because look at how long we've been talking…'

Adrienne checks the clock on her laptop during our Zoom call between Bristol and Newcastle, '…we've been talking for over an hour-and-a-half. It's not a case of, "Oh I eat meat because of this…" It will take me two hours to explain my food choices to you and even then, I don't know for sure that I'm making the right food choices. It's such a complex conversation.'

Adrienne has reservations about rigid dietary identities, believing it's impossible to label foods as simply 'good or bad' for the planet, however well intentioned: 'It's easy for urban people to say, "I am making this food choice based on this X, Y and Z metric," but it doesn't encompass the broader issues. You may choose to not eat beef but what are you replacing that with? The carbon footprint may be smaller, but the socio-political ramifications of that food choice might be a lot worse. It's hard because these are conversations that need to be had.'

I ask why she isn't having them.

'I'm trying to think of the last time I avoided it…' She rubs her chin. 'It was a weekend working with some other environmentalists on a campaign. They were all vegan. I just thought, "This isn't the place to be having this conversation." Although, probably, it's exactly the place where I need to be having this conversation!'

Who are we talking about here?

'It's mostly older, middle-class, white people. Within the university it's young, white – and I emphasise the white – predominantly vegan, at least vegetarian, idealistic people. I don't mean that in the negative sense at all, but you know what I mean when I talk about the young, twentysomething, idealist, environmental activist, anti-capitalist? I would definitely fit into most of those categories.'

Adrienne belongs to a tribe with a strong identity and belief system. So does Ruth, working for one of Britain's most respected conservation charities, and so do I in Bristol, a proud, green city and the first council to declare a climate emergency. But we've all felt slightly on the outside of something, and ever so subtly different to the rest of our tribe. Why? Is it because we understand rural and farming life on a deeper, more personal level than a lot of people in our urban communities? I don't know. But I do know it would be a lot easier to just. . .belong.

I've felt on the outside with farmers and rural folk too. At my parents' ruby wedding anniversary party my cousin Mark wanders over for a chat. I can't imagine him ever leaving the area where we grew up. He loves home. He's a young Millennial and already the ultimate local. He's never been on a train and only flown abroad once for his honeymoon. He went to university in Shropshire and built a successful career as a livestock auctioneer in the county, met a local farmer's daughter, got married, had children, and built a house just two miles over the border in Wales. We catch up and Mark tells me the animal rights protesters have been back, making a noise outside the market.

'I hate vegans,' he says.

'You don't mean that.'

'I do. I hate them. Don't you?'

I'm on the outside again. I try to explain that my vegan friends wouldn't stand outside his place of work screaming at lorry drivers and intimidating people – surely that kind of behaviour has more to do with personal anger issues than your dietary beliefs? My cousin looks doubtful. He's yet to meet a vegan who hasn't shouted at him.

I've felt on the outside when the environment comes up in conversation too.

Back to that day in Iowa, where I'm sitting next to Brandon Pickard in his Case IH combine rolling through a 70-acre field of soybeans. I ask if he agrees that modern agricultural practices have had a negative impact on the environment. He takes some time to think about it and we sit in companiable silence, listening to the quiet rumble of the engine from the comfort of an insulated air-conditioned cab.

'No, I don't agree,' he decides, 'I think we've actually made the environment better. The way we farm now, we raise more bushels to the acre on less nitrogen than what my grandfather did. And it's made the environment better.'

I can't help but form a different opinion based on what I can't see – the things, I feel, are missing from Brandon's farm. There are no trees or hedgerows, no birds or butterflies flying around the combine, no buffer strips along the stream at the bottom of the field. I've pulled over on quiet rural roads across the Midwest, stepped out of my car and strained my ears for birdsong. Nothing. It's like nature is on mute. And so am I.

I've had many similar conversations with farmers around the world – and, mostly, I don't say much back. I am hyperaware of my urban, left-wing media connections and worry that the slightest comment could come off as 'anti-farming' – which isn't who I am. But equally, there is an instinct to challenge, hidden beneath my silence.

Up and down the country, people like me are staying quiet because we feel like we don't belong to either tribe. It's uncomfortable for us. And the lack of challenge and dissension among peer groups is giving a louder voice to those pedalling extreme messages and drivers of division.

Ban red meat! Rewild all the uplands! Ditch cows and replace them with trees! Make everything out of plants! Get rid of farmers and eat lab-grown food!

You may as well declare war on rural communities and shout: Get rid of people's jobs! Get rid of people's homes! Replace one monoculture with another! Erase farm animals like they never existed! Wipe out an entire culture!

The best solutions to our biggest problems are never, ever simple. But simple ideas, like banning this and getting rid of that, are always the most seductive. They grab headlines. I have held my head in my hands at some of the blinkered and uncompromising messages spilling out of the urban environmental activist movement, which are then pounced upon by my own profession – the predominantly urban mainstream media – and packaged up into neat binary debates. On BBC One, Liz Bonin tells me she's stopped eating beef and maybe, if I care about the planet, I should consider doing the same. *Meat: A Threat to Our Planet?* was later found to be in breach of BBC Impartiality guidelines.

On Channel 4, George Monbiot triumphantly proclaims we don't need agriculture at all – not even a wheat field – because we can grow all our food synthetically in labs and factories. (*Apocalypse Cow.* January 2020). In an interview with Radio 4's *Farming Today* he suggests agriculture is entering a 'death spiral' and relishes his parting shot: 'My advice to farmers is to get out now.'

I almost choke on my muesli during an interview on Radio 4's *Today* programme with Hilary McGrady, Director-General of the National Trust, about planting 20 million trees on their land over the next decade and putting 10% of current farmland into woodland. As McGrady explains they will only convert land if the tenant does not want to continue farming it, and no farmer will be forced out, the presenter interrupts: 'But if they're using the land for pasture, if they have cattle, that land needs to be used for tree planting?'

Yuck, cattle. Yeah – chuck them off!

The question is so loaded with piety they may as well have asked: 'What if they're dealing drugs and running prostitutes from the farm?'

McGrady, audibly confused, replies with: 'To be clear, if a farmer wants to continue farming in whatever way – because we don't control how our farmers decide to farm – they can continue to do that.'

Farming livestock is not, in and of itself, a crime against the environment. The decision to eat meat – or not – is deeply personal, based on a myriad of different health, cultural and social factors. It is not simply an environmental question. And are we seriously talking about wiping out farms and farmers, even as we put the food in our mouths, that they grew, that keeps us alive? Are we really OK with that as a national conversation?

The simplistic narrative of vilification has undermined a complex environmental debate and, in my view, dangerously deepened the urban/rural divide. This is not like telling people to fly less or switch to an electric car or insulate their loft or stop buying single use plastic. Food is different. Farming and the rural way of life, as we've already explored, are wrapped up in issues around identity, culture, family, and roots. We're talking about a minority group's entire way of life and reason for being – explosively emotive territory that we should instinctively know demands sensitivity and careful handling, not goading and judging.

'For farmers, it's their identity,' says sociologist Adrienne Attorp. 'Maybe some oil executives are on fire for their industry, but the drivers are different. It's not their father's or grandfather's farm. People say, "oh they'll find work elsewhere", but I'm like – what? Are you going to get a farmer to go and work as an Uber driver or a Deliveroo guy? Firstly, they live in the countryside. Secondly, their entire identity is bound up in raising animals and managing land.'

Only by working together can we tackle the climate emergency – combining the immense power of the millions of consumers in British cities, with the might of those who own and manage 70% of British land. United, what an awesome force for good we could be. Divided, we will all fail. If we don't get our house in order, and fast, this toxic divide will damage our chances of saving our environment.

No more black and white debates. No more stupid slogans and quick fixes. It is time to reveal the truth in all its glorious grey – because the solutions are there for the taking, if we work together.

First, we need both sides to accept the same facts, in the same way. When I travel around the world talking to people about these issues, I'm endlessly shocked by the levels of division I encounter. It genuinely frightens me how the perception of agriculture's impact on the environment can swing from complete denial to all-out blame, with frustratingly little in between. The centre-ground feels a lost and lonely place at times.

Right now, all over the world, farmers and environmentalists are interpreting the same facts to mean different things. Here in the UK this usually boils down to global facts becoming mixed up with national stats.

It's breakfast time on a Saturday morning and there are two pots on the kitchen table. I look from one to the other wondering what to put on my muesli – natural yoghurt or 'oatgurt'? They stack up pretty evenly on calories and nutritional value, even roughly the same amount of calcium.

The oat product has a carbon footprint of 0.54kg CO_2e/kg. This is the 'carbon dioxide equivalent' – a way of measuring and comparing the impact of the goods we buy on the climate. It converts all the different greenhouse gases associated with the production of something, in this case 'oatgurt', into one equivalent amount of carbon dioxide, so a kilo of 'oatgurt' emits 54 grams of carbon dioxide equivalent into the atmosphere. There's no mention of carbon footprint on the yoghurt pot so I look it up. A robust three-year study by the Agriculture and Horticulture Development Board (AHDB), published in 2014, reveals the carbon footprint of a litre of British milk is around 1.25kg CO_2e. That's a lot bigger than the oat product, but well under the global dairy average of 2.9kg CO_2e per litre. I search the yoghurt pot – the milk is British. Phew.

I move on to the ingredients. The yoghurt has one: cow's milk. The 'oatgurt' has 12, including rapeseed oil, potato starch and water. In fact, it's only 12% oats. It also has 'added vitamins and minerals' in the form of calcium carbonate, calcium phosphate, vitamin B12 and potassium iodide, which are all naturally present in milk. Instinctively I warm towards the natural, unprocessed product but then ask myself: does it matter how the vitamins got there so long as they're in there?

But then I swing the other way: do I really want rapeseed oil on my breakfast? And where is the rapeseed oil from? Was it grown with neonicotinoid pesticides, which are harmful to bees?

But then, I suppose, the cows could have been fed rapeseed meal too, so maybe it's better to feed it straight to me and skip the cow?

But what if the dairy product is from a grass-fed herd? That's just sunlight turned into grass, turned into milk, turned into yoghurt. I like the simplicity of that.

But what about the methane the cows burp out? That's a really toxic greenhouse gas.

But methane doesn't linger in the atmosphere for as long as carbon dioxide. And plants take carbon out of the atmosphere, so it's stored in the pasture the cows graze on anyway.

But surely you can store carbon in an oat field too?

But doesn't grassland store more carbon than arable land?

Aaaaaarrrrrrrrgggggggggghhhhhhhhhhhhhhhhh!

I'm hungry and my brain hurts. I lob a dollop of each on my muesli and make an amazing discovery – yoghurt and 'oatgurt' taste great together. I've been mixing them ever since. I eat a metaphor for breakfast.

I'm constantly confused about the environmental impact of my food choices. My gut feeling is that agriculture's impact on the environment has been both overstated and understated. But how can I really know the truth when the same statistics are skewed from all angles to suit every agenda?

I am fed up with feeling split in two and always on the outside. I am tired of ping-ponging between tribes and never feeling like I belong. I am done with sitting on the fence. I want a belief system of my own, that I can hold on to in my urban and rural world. I need a trustworthy answer to one simple question: 'As an ordinary consumer, how do I accurately assess British Agriculture's contribution to climate change?'

I decide to ignore the polarised extremes, and all their obvious agendas, and sniff around the centre ground for individuals I like and trust. I pick two experts – one is a London-based author and environmentalist, the other works with farmers in Devon and Cornwall. What they have in common is expertise, objectivity, fairness, and kindness. Meet Ed Davey and Becky Willson.

I first met Becky in November 2015 when we shared a hotel room together in Belfast. We were attending our first ever Nuffield Farming Conference, along with 17 other newly recruited Nuffield Scholars, all wide-eyed and brimming with enthusiasm.

Becky's chosen study topic was: 'Communicating carbon reduction schemes to farmers; busting preconceptions, driving efficiency and profit'. Back then she was viewed as a bit of a wildcard. Compared to those studying the really hardcore farming stuff, like herbicide resistant arable weeds and the future of the UK potato industry, carbon felt a bit 'out there'. She chuckles at the memory:

'It was something no one was particularly interested in: 'Crazy carbon farming lady – you go and sit in the corner! People couldn't really have a conversation with me because they had no idea what I was doing. Even when I presented my findings in 2017 there was this deathly silence. Conferences used to give me five minutes at the end, now we can fill a whole event just talking about carbon.'

Becky's research, more than anyone else's in our year group, has been catapulted into mainstream collective consciousness, enabling her to span the divide and work alongside farmers and environmentalists. She is technical director of the Farm Carbon Cutting Toolkit, which helps farmers measure and improve their carbon footprint, and is heading up a new Lottery funded project called Farm Net Zero which works with farmers and growers to assess what's possible around reducing emissions and sequestering carbon on-farm. Throughout the Covid lockdowns Becky gave 122 webinars for groups as diverse as village climate action groups (just like the ones my friend Laura set up), secondary schools and farmer groups. She has an amazing ability to combine forthright passion and super-sharp intellect with a down-to-earth humility that immediately puts people at ease. She is a one-woman unifying force for good.

I put my question to Becky: 'As an ordinary consumer, how do I accurately assess British agriculture's contribution to climate change?' She doesn't have a simple answer for me (drat) because there isn't one source of information that applies to all the different production systems that make up UK agriculture. One thing's for sure though, the much-quoted 'agriculture contributes to just under 10% of the UK's greenhouse gases' figure doesn't necessarily reflect what's happening on the ground.

'There's a real dichotomy between what happens on farm and what gets reported in terms of UK greenhouse gas emissions,' she explains. 'So, farmers that are changing their practices on farm, quite often, that doesn't get reflected in these nationally determined contributions.'

That doesn't mean the national statistics are wrong. The National Atmospheric Emissions Inventory, compiled by a range of Government departments, feeds up to the Intergovernmental Panel on Climate Change (IPCC) and is used to set carbon budgets going forward. Becky says these are 'trusted areas to look at' but you need to know what you're looking for and understand how the statistics are compartmentalised:

'One of the big things with these agricultural statistics is that anything positive that's going on within agriculture – so those farms that are producing renewable energy through solar panels and wind turbines and all the rest of it – the positives coming from that don't come off "Agriculture's" carbon footprint – it comes off "Energy's" footprint. If there are any areas of woodland on the farm, that comes out of the "Land Use, Land Use Change and Forestry" category. So, all the emissions are coming from Agriculture's contribution but anything positive that we're doing, like storing carbon in soil and trees, comes out of different pots. Because they've been separated people think, "Well, we can get rid of farming and just have the positives" rather than seeing that they're all integrated.'

Ok, so the bad stuff is obvious while the positive stuff is harder to find. That immediately puts farmers at a disadvantage in public debate. I can understand their frustration with the figures. But Becky also insinuated that the positives relate to 'changing practices on the farm'. My next question is: what needs to change?

For Becky, there are two glaring obstacles standing between British agriculture and Net Zero carbon emissions: synthetic fertilisers and grain-fed livestock. Whichever way she looks at it, they just don't add up.

'These are the areas where we struggle in agriculture,' she says during a Zoom call from her home in rural Devon. 'Why are we feeding grain to ruminants that aren't designed to digest it? That is an inherently inefficient process. It's not an efficient model for the animal in terms of their gut health, it's not efficient for us because we can eat the grain ourselves, and it is not efficient for the climate because it's producing some of those greenhouse gases.'

This doesn't come as a massive shock. I've known it, deep down, for a long time. But still, coming from a Nuffield Farming Scholar, with a degree in Agriculture, who worked on dairy farms for years, it's a bit of a kick in the guts. The truth hurts.

We grain-finish our bulls at home. They spend most of their lives in the shed chowing down on expensive blends of wheat, barley, maize, sugar beet pulp, molasses, palm kernel, sunflower extract and imported genetically modified maize distillers, a by-product of industrial ethanol production. And Dad is so proud of them, these muscly Limousin-Charolais crosses with their sleek coats and shiny noses buried in scoops of grain. They're beautiful boys, all big eyes and long lashes eyeing us over the gate, licking a few escaped flakes of maize from their nostrils. There's only four or five of them in the shed so it hardly constitutes a feedlot – but the system is the same. Dad faithfully feeds his bulls every morning and evening, and I've heard him chatting away to them while forking straw into the pens. Dad and I have had many heart-to-hearts in the shed while feeding the cattle on a winter's night; conversation flowing to the calming sound of munching, hot cow breath clouding the frosty air with that smell I love – sweet hay mixed with the maltiness of milled feed. It was while he was scooping feed from a 600-kilo bag into a bucket that I told Dad I was expecting a baby. The cowshed is a very special place and my heart sinks as I listen to my friend, realisation dawning that our way of raising beef, after all those years of hard work, is environmentally unsustainable.

Could this be the end of the grain-fed beef era?

'I think the era of the ad-lib grain-fed cow is coming to an end,' nods Becky, referring to systems like ours where cattle can munch on grain all day long. 'If we are to have a sustained position for beef and lamb in the UK then I think the amount of grain that's being put into those systems is going to have to be drastically cut if not eliminated, otherwise all our arguments for why you should be eating red meat just fall away.'

And we can't afford to lose those arguments, not least because of the astonishing ecological service provided by grazing animals like cows and sheep. Fields, pastures, meadows, our precious grassland habitats exist because of the ruminant: that amazing creature with the most mind-boggling digestive system, evolved to harvest sunlight by chewing grass and turning nature's energy into milk we can drink, and meat we can eat. And, oh, how they chew that grass! To me the most calming sight in the world is a watching a chilled-out cow lazing on a squashy carpet of green grass and buttercups, languidly chewing their cud, dozing in the sunshine. No animal makes chewing look as luxurious as a cow does. A landscape without them would break my heart.

'It's an indisputable fact that we need livestock in our system,' says Becky, 'We need livestock to manage our grasslands and maintain and restore our habitats. We are never going to get away from that fact.'

The answer, she believes, is grass. Feed a cow grass. It sounds bleeding obvious.

'We can do it,' she says, almost pleadingly, 'We can finish beef off grass and that's a low-cost system that works. We just need to have attention to detail in how we manage our grasslands, our soils, and our animals. In the west of England, where it's warm and wet, what we grow best is grass and we need to harvest that grass as efficiently as possible and turn it into something we can eat.'

Again, this doesn't come as a surprise. I've even asked Dad about it during one of our shed chats: why don't we just raise our beef cattle on grass? It's a lot cheaper than grain and we're in the warm and wet west – surely, it's a no-brainer? His answer is we don't have enough land. Our small acreage of owned and rented ground couldn't grow enough grass or produce enough hay to feed them.

'And you'd never finish these continental breeds on grass anyway,' he says. 'They're too big.'

'Well, why don't we get smaller, native breed cattle like Dexters or Herefords?'

He pats one of his fluffy Charolais calves on the back: 'Because I like these ones. And you have to farm something you like.'

One potential argument in favour of Dad's production system is that cattle fatten quicker on grain than grass. He can finish bulls in as little as 14 months on grain compared to up to 30 months for the heifers, which do spend most of their lives on grass. Put bluntly, the bulls simply don't live as long. That's fewer days on Earth to be belching methane into the atmosphere. I put that thought to Becky:

'There is a certain degree of truth in that, but only if we're looking at it through a methane-only lens. If we're comparing gases, in terms of percentage of emissions coming from agriculture, then methane is much higher than carbon dioxide – and if we get our cattle gone quicker, that percentage goes down. However ...'

I knew there was a 'but' coming.

'...if methane isn't as bad as everyone thinks it is – being a more short-lived gas in the atmosphere than carbon dioxide – then we have to start looking at the emissions going into the animals, via the feed, rather than out of the animals, in the form of methane. It's about separating the biological emissions from the fossil fuel emissions.'New research is also showing improved nutritional quality from grass-fed meat, with higher levels of key nutrients found within this system.'

At 70 years old, Dad isn't going to change his system now, but younger beef farmers will have to make some difficult decisions in their careers.

I pass on exactly what Becky said to Dad, word for word. I'm sitting next to him on the sofa with a mug of tea, reading out loud. I get to the end, close my laptop, and say: 'I guess, just because that's the way it's always been done ...'

Dad looks pensive. He finishes my sentence:

'It doesn't necessarily mean it's right.'

He drains the last of his tea, gets up and walks out to the shed.

I'm feeling a bit bummed out and we haven't even discussed Becky's second point about synthetic fertilisers. At least Dad's in the clear there – we can't afford bagged nitrogen so just rely on cow muck to do the job for us. Turns out, that's the best move for the environment. In fact, Becky can't understand why a ruminant farmer would ever need to apply synthetic fertiliser. She's worked it all out:

'If we've got 78% nitrogen in the atmosphere, we've got clovers which can fix that nitrogen from the air into our sward [pasture] and the fertility source that comes from the animal in the form of their manure, why are we putting additional inputs into that system? This is one of the uncomfortable truths. Nitrogen is one of the most damaging things in terms of its carbon footprint because it is incredibly energy intensive to produce and a by-product of it is nitrous oxide – 298 times more potent than carbon dioxide.'

Becky is growing more animated by the minute, jigging around in her seat in front of the webcam:

'Everyone's talking about methane, but nobody's focusing on nitrous oxide! Everyone's blaming the cows for burping, but nitrous oxide is coming from soils, fertilisers, and manures. How we manage our fertilisers and integrate manure is hugely important.'

I'm totally swept up in Becky's passion and, most importantly, I trust her. But I'm still left with one question – can I eat red meat with a clear conscience? Is my local pub in Bristol right or wrong to take beef and lamb off the menu? Time to call Ed Davey.

I first met Ed in a tent at Castle Howard in Yorkshire. I was moderating a discussion on climate change in front of a live audience at Countryfile Live, the BBC programme's big outdoor event which feels like *Antiques Roadshow* meets Glastonbury meets a Young Farmers rally. It poured with rain and I was sat on some soggy straw bales going through my notes when Ed pitched up, looking flustered. He was soaked to the skin and panting, having ran the last two miles when his taxi from the train station got stuck in the tailbacks outside the main gates. His shoes and trousers were sopping, raindrops plopped off the end of his nose and he had a bright red face from the exertion, but what I remember most was his beaming smile: 'Anna! Hello! I'm so sorry I'm late.'

Ed is high up in the environmental world, serving as International Engagement Director at the Food and Land Use Coalition, a global initiative to transform the world's food and land use systems, and he's Co-Director at the World Resources Institute, a non-profit research organisation. He's a lifelong environmentalist and is on a mission to save the world through optimism. His 2019 book, *Given Half a Chance: Ten Ways to Save The World*, is based around the idea that nature can heal herself, and rapidly, if we just give her the breathing space to do so.

I put my question to Ed over a Microsoft Teams meeting during the 2021 winter lockdown: 'Can I eat red meat with a clear conscience?'

He says the emissions from beef and lamb are high compared to the amount of protein they provide, even if it's grass-fed. Dammit. This is not what I wanted to hear.

'So, we should cut beef out completely then?'

I hope I don't sound huffy. I feel a bit huffy.

'At the very least you could really quite significantly reduce your consumption of beef and, I'm afraid, also lamb.'

I get the sense Ed is treading carefully.

'And that's a great sadness for me and my family,' he adds hurriedly, 'we're having lamb for supper by the way – but lamb is quite high on the list in terms of emissions per gram of meat. I'm not of the view we should cut anything out altogether. I think we should just eat meat sparsely and really high quality. Now, that's easy to say and it's not possible for everyone to do, but I hope that's where we're heading.'

I mull this over. I feel I've worked quite hard in recent years to reduce my meat and dairy consumption. I eat 'oatgurt' for heaven's sake. Compared to some of the rampant carnivores in my family, I'm practically vegan. Why should I eat even less? Surely this doesn't apply to me?

'It's a good question,' says Ed thoughtfully. 'From the point of view of the science, huge swathes of the population in the West and in affluent countries, and affluent sections of societies like the Chinese middle and upper class, the Indian middle and upper class, the Colombian, Argentinian and Brazilian middle classes, of which there are many millions, I'm afraid for that whole group en bloc there need to be very significant reductions. And that's not just folk who may eat meat several times a day, that's all of us who eat meat once a day, or four or five times a week. I'm afraid it probably does need to be less than that.'

He's not finished.

'But there's a very important caveat here . . .'

Oh phew, he's about to give me the good news.

'There are many millions, if not billions, of people in the world who have the right to eat more animal protein. We all have the right, but they in particular have the right and I'm thinking here of Ethiopian families, for example, who have terrible malnutrition and whose children need dairy and chicken and significantly more red meat in their diet.'

Ah. My family do not fall into that category.

'But I'm afraid,' he continues, 'the boundary constraint here is that 25%, or more, of the global greenhouse gas emissions that come from food and land use are largely to do with meat. Alas,' he adds apologetically.

This is all thoroughly depressing for a beef farmer's daughter. Ed writes in his book that: '. . . global agriculture and land use contributes more emissions to the atmosphere than all the world's industry or transport, approximately a third of the world's total greenhouse gas emissions'.

Now, this is the figure that gets farmers hopping mad. While it contains all the emissions associated with agricultural activity, the term 'land use' is incredibly broad. Dropping a heavy acronym now – LULUCF – Land Use, Land Use Change and Forestry, which Becky mentioned earlier. The Committee on Climate Change says it 'covers emissions and removals of greenhouse gases resulting from direct human-induced land use'. That includes forests, grassland, cropland, wetland, and – weirdly – 'settlements' (urbanisation and infrastructure basically). This sector can act as a carbon sink (good), but it can also emit greenhouse gases (bad) through deforestation and the loss of peatlands and wetlands (principally driven by converting land to agriculture), wildfires and degrading natural landscapes for building houses, roads, railway lines, bridges etc. All this wide-ranging stuff comes under 'land use'. It's not just farming basically – but in this context you'd be forgiven for thinking 'land use' only meant agriculture. And while every possible direct and indirect emission from agriculture and its associated activities are included, from cow burps to feed production, only the direct emissions from transport and industry are used for comparison. I can see why farmers get so annoyed by it.

Farmers and growers get very exasperated about shocking global statistics overshadowing more flattering national statistics. Remember the much smaller carbon footprint of a litre of British milk? I ask Ed about it. As consumers, should we be thinking nationally or internationally?

'I think both,' he says, 'We do live in a very interconnected world. We are a trading nation and what we consume here does rely on land use elsewhere and an example of that is British chickens fed on soya from Brazil. That said, assuming that the world population will be 9.6 billion by 2050, there is going to be increasing demand for food. Therefore, where agriculture is productive, we should celebrate that and welcome that, and there's a lot of the UK's agriculture that is among the most productive in the world, so I understand why there's frustration about the conflation. I have a lot of sympathy.'

So, it works both ways. British farmers can't sit in their ivory tower and claim total detachment from the global figures. Imagine if David Attenborough did that? 'The Sumatran rhinoceros is on the brink of extinction but don't worry – we don't have them in the UK anyway.' We are all part of the same global system. Some of Dad's cattle feed comes from South America. Indirectly, our little farm has undoubtedly contributed to deforestation in the Amazon at some point or other. We share responsibility. On the other hand, British farmers and growers are right to shout about their ability to produce affordable food with a much lighter footprint than many other countries. This is a real win for us. We are blessed with a temperate climate, copious amounts of rain, loads of grass and a diverse range of production systems which, though not perfect, give us food all year round and plenty to choose from. Not everywhere in the world is so lucky.

I have a friend who visited a sheep farm in Qatar – one of the most bonkers things she'd ever witnessed. Sheep. In a country of desert. They were importing the grass to feed the sheep and gunning through vast quantities of water. In a desert. Sheep. I mean, come on! Surely leave it to the Cumbrians. That's what Ed Davey means about preserving production in places where it works. (As a sidenote to those British farmers constantly straining at the leash to produce more food – he said preserving, not increasing. This isn't a blank cheque to go hell-for-leather a la post-World War Two. The 'We Need to Feed the 9 Billion' mantra, which was drilled into us as newbie Nuffield Scholars in 2016, has breathed its last from my lips. I'll never say it again. We don't need to 'produce more with less,' we need to maintain with less and stop wasting so much food.)

After talking to Ed and Becky, I can feel something shifting in my own mind. It's an instinctive feeling, taking root as a clear-sighted belief. It is time to lay our cards on the table.

Modern agriculture has committed crimes against its home, the habitat on which it depends – the rural environment. Seventy years of growth, intensification, specialisation, and expansion has damaged the very elements which sustain it: earth, air and water.

The earth, our soils, are still working hard for us but many fields, just like Spratts in Essex, are tired. That naturally occurring organic matter, all those billions of fertility-giving microscopic bacteria, just don't work alongside synthetic fertilisers – a mismatch perhaps only discovered in hindsight – while the soil itself, a farmer's most precious resource, has been allowed to slip through our fingers, either blown off ploughed fields or washed and eroded into rivers and streams.

Vast tracts of land have suffered from an absence of livestock, those chemically treated arable soils thirsting for a dose of nature's fertiliser – beautiful, nutrient-rich animal manure – while, elsewhere, soils are choking on slurry. Saturated pasture and cropland trying to absorb yet another application of slop – like someone struggling to down a pint with rivulets of perfectly good beer running down their chin, wastefully plopping on to the floor.

We are hurtling towards a perfect storm here. As farms get bigger, their slurry pits get fuller and our changing weather gets wetter, the window for safe and legal muck spreading is getting shorter and tighter and scarier. I've witnessed Dad's stress in winter: seeing the cow shed get pooier and the manure heap grow higher, he'll search the forecast for a dry spell:

'I've got to get that muck out,' he says desperately. But the rain keeps on falling and the ground gets more sodden, and the cattle just keep on pooing. He only has 12 cows. I can't imagine the pressure of manure management with a herd of 500 dairy cows, 5,000 pigs or 50,000 chickens.

Meanwhile, those slurry stores, muck heaps and uncovered lagoons silently pollute the air – emitting greenhouse gases into the atmosphere, each contributing to the changing climate which threatens life as we know it on Planet Earth.

Water – precious, life-giving water – is a ticking timebomb if we do not stop polluting it. The criminal dumping of raw sewage by water companies is by far the most shocking and sickening, but the constant drip-drip from agriculture is no better. A slow but constant 'seeping' from leaky slurry stores, farmyard run-off, manure dribbling off wet and compacted fields and fertilisers leaching nitrates and phosphorous into rivers and streams.

Earth, air, water – modern agriculture has damaged them all in some way. It is not an anti-farming conspiracy. It is not a pro-vegan agenda. It is not farmer bashing. It is high time we owned it.

And we are.

Remember – agriculture and nature are intertwined. They can even thrive in each other's company. That unique interconnectedness gives farmers the jump on any other industry to provide the solutions to climate change and biodiversity loss. Just as it has the capacity to harm, agriculture has the potential to heal. That change is happening right now, right under our noses.

Perhaps it's not happening as fast as we'd like, but the change is real, and it's growing, driven by a new generation of farmers who are doing things differently.

Ironically in recent years, as politics and society in general appear to grow ever more polarised, the opposite seems to be happening in agricultural circles. The bitter rivalry and simmering resentment that divided conventional and organic farmers for decades is rapidly paling into history. It's old hat. Today's zeitgeist is the rise of regenerative and conservation agriculture, agroecology, agroforestry, min till and zero till (the practice of leaving soil undisturbed, which basically means no ploughing), pasture-raised, grass-fed and Pasture for Life certified livestock.

You'll find the Soil Association working with plenty of conventional farmers these days. My little project, Just Farmers, brings people together from all walks of farming life. It genuinely fills my heart with joy watching a large-scale vegetable grower and contract pesticide sprayer from Scotland swapping notes on carrot production with an organic market gardener from Devon; or seeing two dairy farmers, one with a fully housed system in Warwickshire, the other running a grass-based herd in Derbyshire having a beer together; or the farmer who raises free-range organic pork from 35 outdoor pigs in South Wales chatting to an intensive indoor producer with 250 sows in a slatted unit in Dorset.

We're led to think all these worlds are divided. But they're not. And they needn't be.

Somehow, we're fed this image of farmers as dyed in the wool, addicts to the status quo, jealously hoarding their precious agrochemicals like

Gollum from *The Lord of the Rings*, but that bears no resemblance to what I see and hear on the ground. Their open-mindedness even takes me by surprise sometimes. In a recent Just Farmers workshop all 12 farmers were glad to see the back of neonicotinoid pesticides and didn't want to lift the ban. And that includes large scale conventional arable growers. I genuinely wasn't expecting that, because I also get taken in by the binary debates in the press or on social media. Just because 'The Industry' says something, doesn't mean individuals on the ground agree.

Leicestershire dairy farmer Ruth Grice, who also works for Nottinghamshire Wildlife Trust, doesn't shy away from the fact there isn't as much birdsong on her farm as there should be:

'I know it,' she says, 'I can feel something is missing when I'm in the fields.'

It worries her; plays on her mind in a way I can't fully appreciate. When you work at a laptop for a living – as I do – it's far too easy to distract yourself from the nature and biodiversity crisis. I can quite easily bury my head in the sand, bathing in the sweet luxury of separation. Problems unseen. Not so for farmers, particularly those with ecological eyes, like Ruth.

They are reminded every day. Ruth knew she had to do something so teamed up with Severn Trent Water to plant 9,000 hedgerow saplings on the farm; a mix of hawthorn, blackthorn, hazel, sweet cherry, dog rose and field maple which will grow fast, thick, and strong, their roots clinging to the soil, holding it in place, while soaking up excess nutrients that could otherwise dribble into the brook, and ultimately into the river. Ruth wanted hedgerow species specifically for the food and nesting habitat they provide for wildlife, calling them the 'unsung heroes' of our countryside.

Change is happening. Even Spratts, that tired old wheatfield in Essex, under the stewardship of George Young, is undergoing the most significant change in a generation.

By the time George's father, Andrew, reached his sixties both he and Spratts, the field which had lost so much life, were in need of a rest.

The time had come to pass the day-to-day management of the farm to the next generation.

Unlike his father, George spent his twenties working in the City of London for a large bank in oil trading. He'd never shown much interest in farming or even displayed much natural talent for it.

'As a kid I was never interested in farming,' he says. 'Until the age of 27, my sole agricultural experiences comprised rolling one field and rowing up linseed straw with the forklift. I placed the heaps so well that when they were burnt, I lit the hedge on fire and the fire brigade needed to come out.'

By 2013, George was fed up with working ridiculously long hours in the City and the complete absence of work/life balance so decided to jack it all in. It seemed a good time to ship out of London and return home to the Essex countryside, and it was a happy accident that he discovered a hidden passion for farming – but not the kind his father knew:

'Before I came back, ecology was not thought about on the farm and I don't think Dad would be particularly offended to hear me say that.'

I first visited George's farm on a roasting hot day during the harvest of 2018. I remember standing in the field next to Spratts, watching the Claas combine moving up and down rows of barley.

In 2021, on a similarly boiling day in June, I visit George again. I am lost for words when I see how much the farm has changed. There, in the exact spot where I stood nearly three years earlier, a tall green crop is on the verge of flowering. Tiny specks of pale blue peep flirtatiously from their delicate buds, ready to burst forth in a blaze of colour. It's a two-acre crop of flax, grown specifically for linen production and will be harvested carefully by hand in a month or so's time. Next to the flax, a green crop of spring wheat shows the first signs of ear emergence, triggered by a spell of intensely hot, dry weather. Barley, which George always hated growing, has disappeared from the rotation completely.

We head down to the lower fields, looking out towards the port where giant cranes punctuate the horizon. Tall grasses and wildflowers tickle my legs as I stride through a waist-high jungle called a 'herbal ley'. I make out some golden-brown blobs in the far corner: cattle. They're new. George's docile herd of native red polls lift their heads and eye us nosily, tails flicking lazily in the hot sun. They weigh up whether we're worth the effort of investigating and, almost with a collective sigh, seem to say, 'Well, we'd better go and say hello.'

They amble towards us, stopping now and then for a munch on a daisy. George is beaming, looking from his cows to me with obvious and palpable pride. Until now I'd always pigeonholed him as your typical east of England arable man – sure, he might have a few cattle on the side, but crops are his thing. Now I see a stockman, incorporating animals into an arable farming system that hasn't seen livestock – or cow pats – for three generations. This deep-rooted herbal ley (basically a thick carpet of grasses, herbs, and wildflowers) will stay on this field for four years, improving soil structure while sustaining the herd. I'm so indoctrinated into thinking of pasture as short grass, it takes me a while to get my head around the fact cattle can eat this stuff. It's so . . . robust. I tug fruitlessly at a chicory plant. Solid.

George checks on his beloved red polls twice a day, sometimes sitting with them if he has time, and moves them on to fresh grass daily, rotating them around 225 acres of herbal leys, in a practice called 'mob grazing' – a core principle of the regenerative agriculture movement.

He rubs the ears of one of his young calves, Albert.

'It's difficult to describe how different the farm feels with livestock on it,' he says. 'There's a feeling of responsibility. I have to be here to care for these animals – it's a tie, but it changes your association with the farm. Before, I always used to see the farm from the yard, or the tractor, so I'd look out at the landfill or the port. Now I see the landscape of my farm from the ground, and it feels like I'm bedded in somehow. You feel a part of it rather than you're just dictating to it. My mental health has got a lot better since I got the cows.'

All this is part of George's transformative farming manifesto – an actual, physical document which he shared online and with his local community under his brand 'Farming George'. It's a bold transition from intensive, conventional production to an agroecological model, which he describes as 'growing a greater diversity of nutritional produce, marketing everything direct, whilst integrating nature into every part of the farming system.'

He is creating a 'wild seam' snaking right through the middle of the farm as a corridor for wildlife. It's basically a massive scrubby hedgerow, from 25 metres up to 100 metres wide, linking up all the wildflower margins, buffer strips and tree belts. This is on top of his agroforestry project. On a cold and rainy February day, George, along with a team of workers, planted 7,000 trees, provided by the Woodland Trust, in a single field.

That field was Spratts.

I feel a strange affinity with Spratts now; it's taken on a personality of its own in my mind. To me, it's the post-war workhorse, the jewel in the crown of the Young's farming enterprise, loyally offering crop, after crop, until, finally, it faltered. Exhausted. Like War Horse.

Spratts is the final stop on our farm tour, and I feel silly for building it up – like it's going to speak to me or something. I'm sure it's just a normal, bog-standard arable field.

Except it's not. It astounds me.

Spratts is also under a four-year herbal ley – which has grown even thicker and taller than the previous field I saw. I lift my knees as high as I can to keep up with George, who's marching on ahead. Lucy, my spaniel, has found her inner springer – fluffy ears pop up momentarily over oxeye daisies and cornflowers before she bounces off again, grass seeds clinging to her coat. The sky sounds full of skylarks – I squint my eyes into the blue, searching – but their chatter seems to reach right to the sun, and I have to look away.

George hands me a stem of dainty pink flowers: 'Persian clover', he says, 'it smells amazing!'

I breathe in the sweet scent, following George towards rows of tiny saplings in the distance.

'I used to feel that my farm wasn't beautiful,' he says, turning round to survey his fields. 'There was very little character or colour, my palette was so small because I was only growing three crops. Now there is so much going on. The changes I've made in the last six months genuinely make me feel 100 times more connected to the farm.'

I trip over a plum tree, almost completely obscured by grass.

'I'm going to have to mow this before Open Farm Sunday,' George muses, adding yet another job to his long to-do list.

We walk the entire 48-acre field passing by oak, aspen, rowan, hornbeam, chestnut, willow, walnut, plum, damson, apple, pear, cobnut, medlar, Mirabelle (a hedgerow plum), almond, gauge, apricot, peach, nectarine, juneberry, cherry, quince and birch, which George is growing for harvesting sugar sap.

'It's going to look so cool in another 30 to 40 years!' he grins excitedly, not caring he'll be an old man by then. Farmers share an enviable peace with the passing of time, not fearing the onset of wrinkles and arthritis but knowing instead there are rewards to look forward to – like sitting in the woodland you created as a young man.

Spratts is the first field on Curtis Farm to undergo organic conversion, which started in September 2020. The agroforestry has also made it considerably smaller – losing 10 acres of productive cropping area. It's brave, some would even say foolish. George has faced a lot of criticism for what he's done – but the former banker defends it with basic maths:

'It wasn't making money before,' he shrugs. 'Even if I get zero harvest from the agroforestry, even if I just farm arable crops around the trees, I'll still be better off so the risk is genuinely low. If I was farming conventionally, it would cost me £10,000 in inputs – seeds, fertilisers and sprays – just for that one field. Now the cost of production is less than £1,000.'

George is convinced Spratts will generate more income in this new agroecological system. He says 40 acres of conventional wheat, at £200 a tonne, would earn him about £24,000. With heritage crops self-milled into flour, he reckons he can make £55,000 to £70,000 on the same area.

'Dad says I have radicalised him when he's out with his friends,' laughs George. 'He understands that the farm isn't making money and that things have got to be done differently but he can't help but remember those halcyon days when you couldn't help but make money.'

Andrew 3, that industrious child of the Green Revolution, has given his blessing but stands on the side-lines, watching his son's agroecological revolutionary zeal from a quiet distance. I detect a note of sadness from George.

'He's been very hands off,' he says. 'I would love him to be more involved to be perfectly honest. I would love to get him more excited about agroforestry. He's happy for me to do it but he's not excited about it. Dad just sees another thing that we're going to have to learn how to market.'

It's easy to paint a rosy picture of George's regenerative journey, but he's chosen a difficult path. His farming career will be harder than his father's. His 2019 crop of lupins, which he wanted to sell as a replacement to soya, failed; most of his willow trees have failed due to weather challenges; some of the other alternative crops, such as buckwheat, have proven difficult to market and sell; and he's spent two frustrating years and £16,000 trying to get planning permission to build a new shed to house his flour mill – a key part of his diversification plans.

'My mill means I'll be able to grow all these different crops,' he says. 'Without this equipment I can't farm as ecologically as I want to. I feel like I'm getting kicked in the teeth at every turn. I've been feeling pretty shitty recently.'

I ask if it would be easier to go back to the fertilisers, chemicals and simple rotations of winter wheat and oilseed rape.

'Absolutely categorically yes,' he jokes, but I detect the genuine note of despondency. 'I sometimes daydream about it. Quite regularly at the moment.'

It's easy to look at agroecology and skip off into the sunset believing we've found the solution to all our problems – but remember how this chapter started: there are no quick fixes. No simple solutions.

Stephen Ware bought into his father's very specialised farming business in Herefordshire. Their two main sources of income were intensive broiler chickens and intensively planted orchards. Stephen, a convert to agroecological thinking, is on a mission to reverse that model and 'extensify' production. He transitioned from broiler production to pullet rearing, which he describes as a 'kindergarten for chicks'. They grow into laying hens and, at 16-weeks-old, they leave Stephen's farm and go to various free-range egg units over the border in Wales.

It's a slower system than churning out fast-growing broiler chickens for meat: 'We're doing two-and-a-half crops a year instead of seven and rearing birds to 1.3kg instead of 4kg.'

On the horticulture side, he used to grow 1,000 tonnes of cider apples for Heineken. The contract went sour during Covid leaving Stephen with a massive excess of cider fruit. Ever the optimist, he decided to use it as a positive ecological opportunity to thin out the orchards, taking out every fifth row and increasing the distance between the trees, giving their roots room to breathe. He's down from 16,500 cider apple trees to 6,000 and has invested in a new 'shake and catch' harvester so he can save the labour associated with hand-picking and market premium apples 'that have never touched the ground' to craft cider producers.

There's more. In 2017, he planted a further 20,000 agroforestry trees, a mixture of apples, pears, and cherries, which he plans to harvest for juicing. He is also pioneering a ground-breaking system in sustainable insect protein production, rearing flies on the by-products of the juicing process and creating a circular farm economy where nothing gets wasted. It all sounds amazing.

But there's a problem. Stephen's agroforestry – the darling of regenerative farming – sprawls across 123 acres of productive arable land. It can grow fantastic crops of wheat, barley, peas, maize, all sorts – except he can't find a contractor willing to faff around all the trees and hedgerows with a gigantic combine harvester:

'The modern agricultural system is built around big machinery on massive areas so if you're trying to do something different you don't stand a chance. The contractors are reluctant to come in because I've got smallish fields with hedgerows, and I've gone and put trees in the way. Some don't even want to talk to me because I'm just that mad bloke down the road who makes life difficult. And they're not the bad guys – they've invested a lot of money in their kit, and they've got to get on, so I understand the pressures.'

'Could you buy your own?' I wonder.

'I can't afford to invest in a combine that sits still for 350 days a year!'

Stephen is caught in the middle – he's too big for the 'niche/smallholder' market, but too small to compete in the dog-eat-dog commodity market. He's also far too busy with the chickens and apples to spend vast amounts of time fiddling around with arable crops that he may not even find a market for. He needs a simple and reliable solution that works.

'Why don't you just let it go wild and take an environmental payment on it?'

'Ha! How much of an environmental payment have you got? I looked at the Environmental Land Management Scheme (ELMS) and it's not formulated yet. The pilot scheme, I felt, is only suited to very large operations or the very disadvantaged operations. For anything in between, like us, which already has hedgerows, and is already operating very environmentally and with some degree of synchronicity with nature, there was very little.'

When we speak in the summer of 2021, he's busy growing wheat, triticale, spring barley and oats but has no idea where he'll sell them or how they'll get harvested.

'I haven't got a plan!' he admits, laughing in that slightly hysterical way. 'It stresses me out more than anything else. I just can't get my head around it, and I really am struggling. Sometimes I just wish I didn't have an enquiring mind, I just wish I would go along with the status quo, I really do. It's just so tough.'

'When I started looking at regenerative farming 10 years ago someone told me: "You really have to keep it at arm's length to an extent, otherwise you'll beat yourself up so much, you'll lose all faith."

I did take heed of that, and I have taken my time with it, but it does keep biting me. That's 10 years ago and our farm does look completely different and it's going in the right direction, but there's far more down moments than up moments.'

Stephen has listened to all the podcasts and been to all the conferences, and felt whipped up with enthusiasm for the agroecological silver bullet – but he quickly discovered its limitations, and started to question the people he calls 'The Evangelists':

'It's very easy to talk about these things but actually doing it is very tough. You've got all the financial burden, all the work, all the stress. The Evangelists can just carry on to the next conference and evangelise there, and the next one and the next one, and it's no work for them, is it? They just turn up and talk about it. We're the ones on the ground trying to make it work.'

It's quickly becoming a hackneyed phrase but it's so true: a farm can't be green if it's in the red. When I read about struggling farming businesses, I wonder how the hell they're going to afford brand new slurry storage to protect our air and water quality or learn the skills to farm regeneratively when they're so focused on keeping their heads above water. You can't just switch at the click of your fingers – that's like someone telling me to forget agricultural journalism and become a financial journalist instead.

I'd be terrified. I'm crap at maths and I don't understand the FTSE 100. This is no different. Farmers will need a lot of help – decades of support to retrain and change – and they won't do it wholeheartedly unless they know the Government and the public are behind them for the long haul.

We have an amazing opportunity to demonstrate that commitment now we've left the European Union and the creaking Common Agricultural Policy (CAP). England, Wales, Scotland and Northern Ireland are developing their first domestic agricultural policies in 40 years. In England, the much trumpeted and eagerly awaited Environmental Land Management Scheme (ELMS) will gradually replace billions of pounds in direct subsidies with payments for what's known in the jargon as 'ecosystem services' or 'public money for public goods'.

Robert the Dorset pig farmer summed up the general feeling towards it: 'I haven't yet met a farmer who has really got to grips with all this ELMS stuff,' he says. 'To announce you are going to take away European subsidies after Brexit but don't replace it with anything the farmers can really understand – we're going to end up being unsubsidised and competing on global markets.'

As the old Basic Payment Scheme dries up and peters out – which for many family farms means the difference between profit and loss – they're not sure what's going to replace it, or even if it will cover the CAP-shaped hole in their bank accounts. If we're to succeed in marrying food production and environmental protection, we need ELMS to work for every farm.

Farmers, however enthused, however determined, face a long uphill struggle, and many will argue they can only change as quickly as the 'system' allows them to. You can't just extrapolate yourself from a powerful farming philosophy that was 75 years in the making; a goliath food system that rules the fortunes of not just farmers, but all the allied industries that also depend on it – processors, retailers, chemical and pharmaceutical giants, machinery dealers, contractors, agronomists.

You may not like it, but they're all interconnected, and they all want to survive. Moving agriculture means moving the whole gigantic beast. You can't just say 'we don't want chemicals anymore' and expect Syngenta, Bayer and BASF to shuffle off and grab their coats. They've got serious skin in the game.

The task ahead of us is gargantuan and the big picture is depressing if you look at it for too long. I'd urge you, in those moments of overwhelmed despair, to resist retreating further into the angry depths of division.

In July 2020, I listened to Tony Juniper, former director of Friends of the Earth and now chair of Natural England, talk about the 'power of partnership' holding the key to Britain's nature recovery. He spoke of the need to create bigger and better-connected habitats across the

UK, enabling us to move from a system of nature protection to nature restoration; and by partnerships he meant landowners and farmers working with conservationists, environmental groups, businesses, Government, and local authorities.

In the past it has always been presented as a binary choice – should we feed people or conserve the birds? And of course, when faced with that choice, the birds have tended to lose out. But Juniper wants to move past the either/or choice and bring nature recovery and food security back together, in a broad, inclusive agenda. He believes we can do the flood prevention, carbon storage, soil health, water quality, biodiversity protection, public access and, yes, food production all in the same place.

And I agree with him wholeheartedly.

But it won't happen unless there is a mainstream culture shift towards the middle-ground in this ideologically divided space. A place where both 'sides' concede they will have to make sacrifices and seek compromise.

Hardline environmentalists are never going to achieve their rewilded utopia, untouched by the hand of human industry, in a modern world with a growing population and billions of mouths to feed. Hardline farmers are never going to retain their isolated islands of intensive agriculture, untouched by the hand of environmentalism, in a modern world where cows are contentious, and consumers want more than just calories from the farmed landscape.

A rude awakening is urgently needed on both sides of the environmental debate. It's time to ask ourselves what we're really fighting against. Is it climate change or is it each other? The rage of the left-wing activist who feels no empathy or respect for the Tory factory farmer? The fury of the Cumbrian hill farmer who looks down upon the idiocy of the clueless London vegan? How much are we even listening to the message on the flag, but judging the person holding it? We need to wake up from daydreams of what we think should happen and get on with what actually could happen.

When you feel lost and depressed at the state of our world, try turning your eyes towards the little people, inside the big picture. They're there if you look hard enough.

There's George Young giving large parts of his farm back to nature, walking through fields of crimson clover, trailed by Albert the calf. There's Becky Willson going from farm to farm showing farmers how to measure their carbon footprint. She points out their problems but shows them their proudest environmental achievements too, and sometimes it's the things they'd never even noticed.

When Becky visits our farm to meet Dad, she pushes a spade into the ground on The Hill and digs up a small piece of topsoil. She holds it in her hands and exclaims at all the worms wriggling around: our small piece of earth bursting with organic matter. Dad looks on, surprised and proud, with a smile on his face, 'Well, I've never really stopped to think about what's under the grass,' he says shyly, 'but now, come to think of it, when I'm walking the fields on an early morning, I can hear the worms.'

I've never heard him say this before.

'Yes,' he nods, 'I can hear them sucking back into their holes on a dewy morning.'

It proves the crucial point – very few farms are all good or all bad. Yes, our grain-fed cattle are a source of emissions – but our soils are packed with carbon, a sink. It doesn't mean we're net zero, we're not, but it does mean Dad has a positive contribution to make. Our pasture, our soils, our worms – we're doing our bit.

Becky looks at the big bags of grain in the shed – a system she passionately believes is inefficient and unsustainable but speaks to Dad with sensitivity and understanding. And he listens to her with respect, and admiration for her knowledge and specialism. She makes an impression. He talks a lot more about manure management since meeting Becky.

These kind of open-minded conversations between the 'Fergs' and the 'Greenies' are happening up and down the country, hidden from the media spotlight and a million miles from the shouty arena of polarised public debate.

My sister Kate is a conservation officer for Buglife in Shropshire. She's working with landowners and local councils to create a 'pollinator highway', creating and restoring 20 hectares of wildflower-rich arable margins and grasslands throughout the county.

Kate talks to farmers and landowners every day – she is on the frontline of all these issues – and yet the shouty, combative world she reads about in the papers, or hears on the news, bears no resemblance to what she experiences on the ground:

'Ultimately a story about a bad farmer gets more reaction than a story about a group of farmers planting hedgerows,' she says. 'Conservationists and farmers are working side-by-side on the ground, and both parties understand that they need the other. Of course, there are difficulties, just like any other partnership, but largely I experience amicable and respectful relationships.'

When I listen to the people in the centre ground, when I see farmers and conservationists working together on a riverbank – quietly doing their bit without debate or drama – this is what motivates me to save our planet. This is what rouses my revolutionary spirit to act on climate change. Because it reminds me that we can all make change happen, however small.

It isn't the cows or the bees or the trees standing in our way. Prejudice, division, and tribalism is getting in the way of climate action. We can solve our problems tomorrow if we do it together. And when we meet in the messy middle, we might realise we're not so different after all.

Aldo Leopold was an American ecologist and conservationist who wrote *A Sand County Almanac* about the flora and fauna around his Wisconsin home, published in 1949. He wrote beautifully about the natural world. I'd never heard of him until I attended the South Dakota Cattlemen's Convention in 2017 when, below a pensive photograph of a rancher surveying a prairie of swaying grasses, were the words: 'The landscape of any farm is the owner's portrait of himself.'

The second time I heard Aldo Leopold quoted was during a conversation with my sister Kate when she was trying to describe the sense of grief she feels when observing the environmental degradation of our natural landscapes: 'One of the penalties of an ecological education is that one lives alone in a world of wounds.'

When a cattleman and a conversationist choose the same author to articulate their feelings about the world around them, they can't be that far apart. Their experience cannot be so different when they are moved by the same words.

For me personally, the most stirring ecological experiences of my lifetime have been in landscapes where farming and nature sing together, in perfect harmony. On a working farm near Wells in Somerset there are wildflower meadows. Nestled without fanfare at the foot of the Mendip Hills just off the A39, hidden behind hedgerows lined with cow parsley and dandelions, lies Chancellor's Farm.

It is one of those perfect days, as spring eases seamlessly into summer, when I'm met with a breathtaking vista of blue and yellow. 'Chelsea colours,' as my footie-loving father would say. I look closer. I'm sure they're bluebells but doubt myself – aren't they a shade-loving woodland flower? I check. They are bluebells – out and proud in the sunshine with the vibrant yellows of tormentil and common and meadow buttercup.

It sings to me, in the same way the meadows of Transylvania sang to my ancient soul, my primal humanity. Nature is here, waiting patiently in the wings for us to share the stage with her once again. And her partner hasn't forgotten her. Though he's hogged the limelight for a little too long, Old Macdonald is realising the show is much better with his leading lady, Mother Nature, by his side.

CHAPTER EIGHT

COMMUNITY

J ANUARY 2021 WAS DARKER THAN most years. There was no better backdrop for a nation in the grip of illness and death than the black nights and dull days of midwinter. As the second wave of Coronavirus ripped through our people, claiming thousands of lives, I lost my baby.

The pregnancy, my first, ended as suddenly and surprisingly as it had started. A little life sparkled for a moment and then, for reasons only nature knows, they had to go. I was left with the strangest ache – a mixture of love and gratitude for the journey we had shared, however short, combined with a sadness that knocked the stuffing out of me. Even now, if the tears decide to come, I am powerless to stop them.

In a small room at the Early Pregnancy Unit in Bristol's Southmead Hospital, I lie on the bed as the friendly sonographer busies around preparing the scanning equipment. I don't quite catch her name so check the tag on her NHS scrubs – Carrie Montgomery – and though her face is largely concealed behind a face mask, I can tell she's smiley. I warm to her instantly and conversation flows as freely as if we were sitting in the hairdressers.

Carrie asks what I do for a living and before I know it, I'm telling her about my book. Perhaps it comforts me to detach momentarily from what's really happening, so I witter on about the urban/rural divide:

'It must sound very niche.' I peter out, suddenly self-conscious.

'No, not at all! It's fascinating. I'm definitely a town girl,' says Carrie. 'You should talk to my friend Katie. She moved to the countryside from London and had a tough time with it. I'll put you in touch if you like?'

And that's how I came to meet Oxfordshire-based writer, yoga teacher, reiki practitioner, and wellbeing coach Katie Jones. We talk over Zoom, just a week or so after the publication of her first book *And Then She Woke Up: How to RESTORY Your Life*, a semi-autobiographical guide to dealing with childhood trauma, inspired by her own shocking realisation, at the age of 41, that she had been abused as a child, pain she had blocked out and buried for most of her life.

Katie grew up in Neath, South Wales – which she now describes as 'a depressed place' where everyone seemed to be struggling:

'It had that real Valleys mentality of "life is hard". It just all felt a bit grey. There wasn't much colour, light or life.'

Katie left as soon as she had the chance, to study Spanish and Italian at university in Cardiff.

'Not through choice,' she hastens to add. 'I applied to Edinburgh because I wanted to get as far away as possible but, sadly, they didn't accept me, so I ended up in Cardiff. I hated it. I think it was too close to home.'

She spent her third year living and studying in Italy and Spain, worked in Majorca for a while after graduating, and then moved to London:

'There was a definite sense of bright lights and adventure. The prospect of going back to Neath wasn't even an option. Even if I had no money and no job, I couldn't go back to that depressed way of living.'

London did not disappoint. She took to it instantly. Katie met her now husband, Richard, and they spent three years exploring the city together. In hindsight, it wasn't long enough:

'We ended up relocating to rural Oxfordshire way sooner than any of our peers. Everyone was like, "What are you doing? You must be mad!" and on reflection I think we were. We still had so much more living to do in the city. Life was good. My husband had a flat in Notting Hill, we had excellent bars and restaurants on our doorstep and the buzz going down Portobello Market was what I had been looking for – the excitement.'

'Why did you leave?'

Katie groans.

'It was work. My husband, then my boyfriend, had an opportunity.

We were very young – not even 25 – and we moved into this sleepy little Oxfordshire village that had no shop, one pub and a church. It was a huge culture shock and such a departure from the bustling city. We moved in the December and in January, there was a Burns Night supper in the village hall. It was the first social thing that we'd seen advertised so we were quite excited.

'We walked in and the whole place just went quiet. We were by far the youngest people there. Everyone was very friendly but in a grandparent kind of way. At the end of the night, I just remember saying, "What have we done? This is our social life now."

'The pub back then was like tumbleweed, there was nobody in there and there was no atmosphere. Friends would come and visit us from London and ask, "What do you do? Where is your fun?" I genuinely don't know what we did. There really was nothing. It was tough.'

The village came into its own several years later when Katie and Richard got married and started a family. It put a stop to the gossip about the young couple 'living in sin' (Katie insists this was a thing. She remembers going to a drinks party and putting a ring on her wedding finger, so she didn't have to face conversations about being unmarried.) Having children opened up new opportunities for socialising. Neighbours she never knew existed started popping up at mother and baby groups and outside the school gates:

'You realise these people were there all the time, but your paths haven't crossed because you're off to work and they're off to work. Suddenly this quiet little village became the "chocolate box" dream and we could walk to the village school and there were thatched cottages, and the children had their friends, and it was just the most beautiful place to raise a family.'

Fun-starved Katie hungrily watered the seedlings of friendship and built a new social circle. There were dinner parties left, right and centre – tipsy parents stumbling out of each other's houses every weekend – and she threw herself into community life, saying yes to everything:

'I was an avid people pleaser – I'm not now – but back then I was desperate for approval. I got massively involved with the village school, I worked in a number of voluntary roles, and we were always off to a fete or something. I was super pleasant and helpful, and people would know: "Go to Katie and she'll say yes." I desperately wanted to fit in and to have people like me, but then it reached a point where resentment built up and I wondered, "Why am I doing all these things I don't want to do?" Suddenly, it became very claustrophobic.'

'But at least you were welcomed into the community?' I suggest. 'Lots of people would love that.'

Katie isn't so sure:

'I would ask yourself the question: do you really want to be part of it? It's quite possible they're never going to see you in the same light, you'll never be one of them. You can bend over backwards trying to please people when you're never going to fit in anyway. My advice to my younger self would be: just be yourself and do what you want to do. If you want to get involved, get involved with activities that bring you joy but don't feel obliged to do it all. If you dislike bell ringing, don't sign up for the bell ringing – even if everybody's looking at you to volunteer. Often, they'll think, "Ooh, there's new blood in the village – great! We can give them that job and that job." I ended up doing all sorts of random stuff which, after a while, wore me down. I was giving so much of myself to others that there was nothing left to give my own family, which was back to front.'

In a complete departure from the naïve Londoners who'd pitched up in rural Oxfordshire in their early twenties, desperate for friends and community, Katie and Richard found themselves, more than a decade later, yearning for more isolation. They moved house again – this time to a smaller, quieter hamlet.

'Nobody really knows anyone here and we love that sense of anonymity,' says Katie. 'In the previous village everybody knew everybody's business. When we moved here, I made the decision I wasn't going to volunteer for everything and has it affected my enjoyment? Not at all.'

'Really? You haven't felt on the outside?'

'No. I have now learned that my happiness is my responsibility and comes from within. I think I spent decades searching for acceptance in my surroundings, because I was so unhappy with what was going on inside me. It is perfectly possible to exist happily in a village or in a rural environment and not be fully accepted by the people who have lived there all their lives.'

Katie enjoys the isolation so much, she's ready to move again – even further into the sticks:

'A house we looked at recently had no neighbours at all. It was really, really rural.'

'That doesn't intimidate you?'

'Not at all. But then, I love nature. Part of my healing journey has been to spend time in nature and soak up the peace. It's difficult to do that when someone is playing loud music or there are sirens going past. I think having complete quiet and green all around you is really important ...'

She checks her privilege.

'I know you can't always achieve it but, if you can...for us, it does appeal.'

Those who can achieve 'complete quiet and green' are a tiny minority. The average price for a property in Oxfordshire, at time of writing, was £463,336.

There's a huge push for development right across the nation – town and country – propelled by the Government's desire (though no longer a target) to deliver 300,000 new homes a year in England. The battles rage in council meetings all over the UK, and with intense ferocity in the southern counties of England. The most entrenched planning wars make the now-infamous Jackie Weaver Zoom call, which lit up social media in February 2021, look like a bit of friendly banter. Friends can easily turn into enemies when communities fight for – or against – development in their towns, villages, and hamlets.

Wiltshire dairy farmer Peter Gantlett sat on Clyffe Pypard Parish Council from 2000 to 2021 and was its chairman for 16 years. Alongside his wife, Anita, they made scenery for children's pantomimes, planted trees outside the village hall, washed and oiled the floor inside the village hall, made teas for the plant fair and blocked off a broken stained-glass window in the church to deter a parliament of rooks from filling the clock tower with nesting material. The parish, which includes the village of Bushton, is tiny and very rural. It's home to 300 people living in about 150 houses, has six farms (four of them council owned), a church and the much-loved village hall. Both the pubs have shut down and there is no school or shop.

The Parish Council gets together four times a year and Peter never missed a single meeting in two decades, which adds up to about 300 hours discussing parish business. Around 2015, he found himself on the frontline of the innocuous-sounding National Planning Policy Framework, which sets out the Government's planning priorities in England and provides a way for local authorities to prepare their own plans for housing and development. Wiltshire was selected as a pilot area for Neighbourhood Planning, which aims to give local people a say in what gets built and where.

'It was an opportunity for us to create a mass of opinion that couldn't be ignored further up the chain,' says Peter.

He was keen to find out what the community thought so two questionnaires and an official housing survey were carried out. According to Peter, 'in all three, the majority of people were accepting and supportive of some development.' The housing survey, conducted by Wiltshire County Council, identified a need for five affordable homes in the parish.

'Our community is struggling, and our housing stock is getting more expensive,' he says. 'There used to be a youth group in the village and while most of the parents are still living here, very few of the children are. They've moved away. If we don't allow some development the community is going to wither away.'

Peter was keen to make some room in the Neighbourhood Plan for affordable housing, which made him very unpopular among several people in the community:

'It was nimbyism at its extreme,' he says, wheeling out the acronym that crops up in every planning dispute: "Not In My Back Yard".

'People move into the parish, get their little house in the countryside and then they want to pull up the drawbridge and not have any more development. Where do they think their house came from in the first place? When you think of the urban/rural divide, people assume it's the cities against the countryside but it's not – it's within the communities. People come here with an urban mentality, wanting to live this rural idyll but not realising that the rural idyll doesn't exist. They want to see cows grazing in the fields but they're only there because someone milks them every day, and a lot of people who milk cows in our parish live in mobile homes. Is that sustainable? Shouldn't those people be housed in nice affordable homes?'

Now you could argue that Peter, whose rather lovely house is surrounded by a thousand acres of his own farmland, holds all the cards here. He lives in an impregnable fortress, a green moat of security, which protects him from having to look at ugly new houses and developers can't touch him unless, of course, he decides to sell them some land to build on. His 'pro-development' overtones made him a few enemies and things got ugly:

'I was chairman of the parish council and the driving force behind the steering group – I was the person pushing the Neighbourhood Plan – so I think they wanted to side-line me and shut me up.'

Two separate complaints were made against Peter, which were referred up to the Standards Assessment Sub-Committee at Wiltshire Council. One accused him of breaching the code of conduct by not declaring a financial interest or properly disclosing his intentions to provide land for development (something he would be willing to do but more 'as an act of altruism, because land for affordable housing doesn't attract a prime premium'). The second complaint hurt him the most – an allegation of bullying and intimidation.

After a lengthy process of investigations and an all-day hearing, it was concluded that Peter had not breached the council's code of conduct, on any account. No further action was taken. It took more than a year to resolve, and in recalling it, I can tell he's still upset.

I've known Peter for several years, we met through my communications project Just Farmers, so we trust each other. I doubt he would have shared his story with me otherwise. He is calm, rational, and almost frustratingly measured – you can't rush or rattle him – so I'm shocked to see him so shaken. The ordeal took a serious toll on his mental health:

'I've never really had much time for all this mental health stuff, but it made me appreciate it for the first time. To be falsely accused of something is a serious thing and the complaints process is brutal. It knocked my confidence a lot. I'm not the sort of person to start crying about things, but the emotion took over really. It was like standing in front of a steamroller and slowly being squashed, and no one would apply the handbrake.'

Peter, at the behest of his loyal parish clerk, reluctantly agreed to run in the parish council elections of May 2021 but lost.

'I am so relieved to be out of it,' he smiles for the first time, like a weight has been lifted. 'I would have felt guilty if I hadn't run, like I'd abandoned the community, but at least now I can move on with a clear conscience and draw a line under it. The Neighbourhood Plan is still going ahead and I'm happy about that. I do think development is almost inevitable because we have a need for it, but it's left me very battered and bruised.'

I mention Peter's experience to Katie Jones in Oxfordshire, and she comes straight out with it:

'If I'm honest I'm probably a NIMBY. We've made a decision to move out of the city and into a village, now a hamlet and potentially somewhere even more remote because we're seeking the peace and quiet; we're seeking the green space. We bought our house, which looks out over a paddock, specifically for that reason. Therefore, if something out of our control changes that environment for us, well, I'm not going to be happy, am I? The thing we value more than anything is not having our peace disturbed. If suddenly a huge housing estate was to be built, then I'd be really sad. It's easy to say I support more rural housing, but would I? If it was in my backyard? No, I probably wouldn't.'

I find Katie's candour enormously refreshing and I respect her for it. She's voicing what thousands think in private – and that takes guts.

Tensions around who gets to live in our countryside will inevitably deepen as more people seek to leave the cities post-Covid and start new lives in rural areas. At first, I was sceptical as to how real this trend actually is – it feels far too convenient as a news story, exactly the sort of neat little press release we journalists drool over. Lockdown Londoners escaping to the country? Brilliant! Pretty pictures, human interest, and likely to spark some debate too. It ticked a lot of boxes.

But it turned out to be true. A Rightmove survey of more than 4,000 home-movers in May 2020 found almost one third of buyers said lockdown had made them want to live in a rural area, and one in five first-time buyers also said they wanted to move to the countryside. It would have been easy to write it off as a springtime blip, driven by the novelty of lockdown, but the trend appears to be holding.

Small towns and villages within commutable distance of London are top of the shopping list (Katie and Peter's stomping grounds would fit into this category), plus spots by the seaside.

According to the Rightmove survey, buyers are driving the city exodus phenomenon – only 13% of renters said they wanted to live in a rural area – but we are two of them.

In April 2021 Alex and I, tired of our 'Bristol or Shropshire?' six-year stalemate, decided to give my homeland a go. We rented out our spare room in Bristol to some friends (a couple who had been living with their parents in lockdown and going gently stir crazy) and we rented a little house in Shrewsbury, as a trial to see if the dream really could work in reality.

It would be a stretch to say we've 'escaped to the country' – we're a stone's throw from the town centre – but we find ourselves in one of Britain's most rural counties, only a 30-minute drive from my parents' farm and I can already feel my roots reattaching. It's the simple things – recognising place names on signposts, discovering shops and cafés I knew as a child are still open, and most wonderfully of all – hearing birdsong from my bedroom window.

I can see trees from the lounge and hear church bells on a Sunday. For the first time in my life, I'm gardening. I grew a dahlia. Alex, like a closed bud, is opening up gradually, with each new discovery – a good running route, several allotments, a speakeasy cocktail bar, hipster pubs serving craft ales, and an invitation to join a local five-a-side team. Slowly he's realising that the things we love doing are not limited to the city, but Bristol, I know, continues to hold his heart.

We've decided to give it a year before making a decision. Maybe we'll spin a laminated card with BRISTOL and SHREWSBURY written on either side, like the people on *Wanted Down Under*.

I know we're not the only ones.

'It is definitely something we have seen in a rise in, especially in the last year or so,' says our super friendly letting agent, George Hyne, who works at Belvoir Estate Agents in Shrewsbury.

'Any sort of four-bed property we put on tends to have two or three minimum applications from city folk wanting to escape. For £1,400pcm they can find a large farmhouse with a garden, rather than a two-bed flat in the city. Will the trend stick? I don't know. Lots move up with the idea of renting for a couple of years prior to buying so only time will tell if they stay.'

For the record, we're renting a modest two-bed, but I'm well aware of how lucky we are to be able to do this. A day doesn't go by when I don't feel thankful for the opportunity to 'try before we buy', and for having a partner willing to share in the adventure.

Sometimes I even feel a bit guilty. When you've spent 15 years in rural affairs journalism, covering issues like the lack of affordable housing, the impact of second homes and holiday cottages on rural towns and villages, low incomes and poor public services, farmers tearing their hair out with inconsiderate dog walkers and posting pictures on social media of horrific attacks on their sheep, increased littering, trespass and anti-social behaviour in beauty spots over lockdown – all this combines to make me deeply self-conscious as a city dweller. I automatically feel tarnished by my postcode. Just as our submissive spaniel Lucy crawls along the ground on her belly, rolling over for every dog who passes by, I bow and scrape beyond the city limits, nervously scuttling around the countryside like Manuel from Fawlty Towers, terrified of cocking up.

It's probably because I'm used to hearing first-hand how some 'locals' feel about 'incomers' – it's a frequent topic of conversation among my rural friends and family. Someone I know very well can't help but get frustrated with the friendly London surgeon who bought the big house down the lane as a second home. He chainsaws a lot because he finds it therapeutic after a stressful week in the city.

'He doesn't consider us yokels who have to listen to his blooming chainsaw all weekend,' she grumbles. But she won't say anything to him – she's too shy.

I talk to sociologist Ruth McAreavey about it. She grew up in 'deeply socially conservative' rural Armagh in Northern Ireland and has dedicated her academic career to the study of rural communities.

'Your friend is railing against gentrification, and I understand that,' she says, with genuine empathy.

Ruth is all for more people moving to the countryside. In her mind it has the potential to solve a lot of problems – easing population pressure in urban areas, attracting money to underinvested rural areas, reinvigorating struggling communities and generally creating a more diverse, bustling rural economy:

'I do a lot of work with Europeans, and in Spain rural communities are dying on their feet because of depopulation. Whereas in rural England it's much more a question of gentrification and that concerns me.'

I tell Ruth about my childhood home – our tiny hamlet on the Shropshire/Welsh border. In the early 1990s, it was all 'locals' (Nan and Grandad next door, Mrs Lewis my primary school music teacher up the road and our late and beloved friend 'Little Geoff', a bachelor farmer who lived down the lane). By the time Little Geoff passed away in December 2014, it was almost all 'incomers'. Mum and Dad's farm now has a second home or a holiday let on every side. Ruth doesn't even look surprised. It's a familiar tale to her.

Ruth is pro-people but anti-exclusivity. She worries greatly about city money driving out local working people and turning the countryside into a playground for the well-off:

'I think that's a real danger, when we put so much weight on the exclusive rural space, which has happened during the pandemic, people want it more and more. If it becomes so gentrified and elite, it creates a very monochrome society. I know housing scheme agreements are there to try and get that social stratification into new-builds, so you get the executive homes and starter homes, and I think that needs to happen more and more. Otherwise, what you've just explained, will be the case for most villages.'

'You should talk to Sally Shortall,' she adds thoughtfully, 'she's done a lot of interesting work in this area.'

Professor Sally Shortall is based at Newcastle University's Centre for Rural Economy. She is president of the International Rural Sociology Association and has advised the Organisation for Economic Co-operation and Development (OECD) on rural-urban interactions. She is a leading expert on rural development and rural proofing (the process of reviewing Government policies to ensure equitable treatment of countryside communities).

Sally, a farmer's daughter, takes a tough love approach to her subject. She hates it when rural areas are presented as a poor underdog. This assumption of rural disadvantage in comparison to urban privilege, Sally believes, is the 'folly of rural proofing' and overlooks the real-world complexities.

She pops up on the Zoom call looking every inch the academic. Long shelves packed with books stretch the length of her office, way beyond the webcam. She connects to audio and a strident Irish accent booms from my speakers:

'They're so privileged, they're so wealthy, they're so elite, they're so exclusive, they're so old and they're so white – because nobody else can afford to live in them!'

She does not mince her words about Britain's most desirable rural postcodes.

'And then you have some of the ex-mining communities, that are also rural, and it's practically impossible to give a house away because nobody wants to live there!'

Sally grew up on a working farm with her six brothers in County Laois about 60 miles south of Dublin. Her mother was widowed at my age with seven children all under the age of 15.

Rural Ireland, with its plethora of small working family farms and socially mixed communities, bears little resemblance to the large privately-owned estates and exclusive commuter villages of rural England. Sally takes an objective view, she's not ruled by sentimentality, which helps her think outside the box:

'If you look at East County Durham, there are horrendously poor ex-mining rural communities there. West County Durham is quite wealthy – Barnard Castle and places like that. I think there has to be something targeted at those rural areas that are struggling – maybe you could invest a lot in the buildings there and try to sell it to urban people who want to move post-Covid? Try and make it an attractive rural location so we're not just focusing on Keswick and Alnwick and the Yorkshire Dales. I think all of rural England needs to be opened up.'

I'm nodding along enthusiastically. I'm in. Where do I sign? But then Sally backtracks, taking the wind out of her own sails:

'But it's the power of this lobby group,' she sighs, naming no names but referring to some giant omniscient collective of NIMBYS.

'They have a lot of resources. They've got influential people with a lot of power who want it to stay as it is and protect their privilege.'

I play around with some hypotheticals:

'How would it go down if there was a concerted campaign to fill the countryside up a bit? Encourage more ordinary urban people to move there?'

Sally half laughs, half scoffs: 'We already see how they react!'

She ticks off some local examples:

'Any attempt to build houses in Almouth or Alnwick and you have huge campaigns against it, and then they'll moan about the loss of services and schools closing. I find it a really interesting dilemma and a peculiarly English one. You don't have this in other parts of Europe.'

England dominates our conversation because Ruth and Sally are both based in Northumberland. It's their point of reference. All the same planning and housing tensions exist in Scotland, Wales, and Northern Ireland (have you seen Snowdonia and the Highlands at the height of tourist season?) But with 84% of the UK population crammed into England, it's an inescapable fact that the gears are crunching louder in the Anglo corner. Just looking at the population statistics for the South East makes me feel claustrophobic. On one of the hottest days of the year, in June 2021, I got stuck in traffic on the Dartford Crossing, sandwiched between two articulated lorries in a tiny Peugeot 207 with no air conditioning. The traffic, the noise, the sheer scale of the buildings and infrastructure in every direction – it was panic attack territory.

It's mad that most of us are stuck in one expensive corner of the country. Why can't we spread the love a bit?

Listening to Sally, I find myself getting annoyed. She's right – we can't have it both ways. If I object to an affordable rural housing development, which will bring young families into my community, I can't then turn around and moan about crap public services and the village dying.

You can't have a real community without the people. But you can have imagined communities – and that's what the political scientist Benedict Anderson wrote about in his 1983 book, which explores the origins of nationalism. He says we construct ideas of community, imagined by the people who perceive themselves as part of that group, except the reality often doesn't live up to the fantasy. Affordable homes are not thatched cottages. It doesn't fit with the dream. We fall in love with the idea of ruralism.

In 2011 the OECD published some research on rural England – and this is where Sally has a real mic drop moment:

'By the OECD's classifications – England isn't rural. It flips between England and the Netherlands as to which one is the most densely populated country in Europe, but because of the psychological and cultural commitment to the rural idyll and rural way of life in

England, they agreed to study it rather than rule it out on the basis of their classifications.'

I'm astounded. We think ourselves rural. To the outside world, England is a postage stamp with 56 million people squashed on top of it, but something in the bones of an English countryperson is wedded to an almost spiritual belief in our green and pleasant land. The religion of ruralism.

My conversation with Sally Shortall blew my mind a little bit.

If the mass post-Covid relocation transpires — more 'townies' moving to the country — there will be cultural clashes. On top of the measurable impacts on house prices, property availability, not to mention the effect of increased homeworking on the fortunes of our city centres, there will be soft ripples too — the human stuff. Unspoken tensions, undercurrents of discontent, mumbling and grumbling in the villages and hamlets. It's already there and has been for years.

In January 2021, right after the miscarriage, Alex and I hunker down for lockdown in a holiday cottage in deeply rural north Shropshire. It has been booked for months and the owners, thankfully, still allow us to come, in spite of national restrictions, because we're using it as a place of work. It's the warm-up to a bigger move, a let's-see-how-we-get-on-for-a-month kind of deal.

On a dreary afternoon I take Lucy for a walk. We set off up a rather busy B-road where every passing vehicle seems moody and in a hurry. They're not, it just feels that way when you jump on to the slither of verge (there's no pavement), yank the dog out of the way and squeeze yourself flat against the hedge. A pick-up and stock trailer roars past — carrying a whiff of sheep poo on the wind. Moist specks of mud spatter my face. We press on, the sky darkening as a grey listless day, which never really got going, decides it can't be arsed and hands over to dusk. I haven't got a map. I made the decision to busk it, trusting we'll come across a footpath at some point. I'm sure I passed a brown tourist sign in the car earlier. Maybe there's a country park or some nice woodland around here?

Sure enough, we come to a little wooden signpost — a bridleway leading off the main road. Grand job. It takes us through a quiet empty farmyard, though the lights are on in the cow shed. I hurry past, feeling like I'm trespassing (even though I'm not). A kissing gate goes into a sheep field. I put Lucy straight on the lead. We follow the signs onto a crop of turnips being strip-grazed by store lambs – agriculturally productive but aesthetically very displeasing – and I teeter along the electric fence line trying to step on the green bits to avoid the cloddy, clagging mud. We cross several more fields – pasture now – and the signs run out. Great. A crow caws – a lonely, desolate sound to break the silence of deep winter. Even Lucy looks miserable; doleful brown eyes pleading: 'I'd rather be in Bristol.'

Et tu, Lucy?

It's virtually dark, there's no obvious path and I'm imprisoned by hedgerows and electric fencing. I can't even see a gate. I hear the rumble of an engine coming closer – a quad bike, no doubt the farmer doing their evening rounds. Fab. I'll flag them down, ask for directions and we can get the hell out of here. Except the person on the quad bike completely ignores me. I wave my arms and shout 'hello!' but they turn around and drive off. Surely, they must have seen me?

I retrace my steps and trudge back to the pongy turnips. I look down and take stock of my appearance and start chuckling. I'm wearing my Joules tweed coat, brown knee-high leather boots, a bobble hat, and I have a spaniel in a cute tartan harness, complete with a tiny bow tie, on a lead. I'm standing in the middle of a field of turnips, caked in mud, in the dark, surrounded by sheep. I laugh and laugh. It's like someone plucked me out of *Country Life* magazine and deposited me in the most inappropriate, unphotogenic rural scene ever. What must I have looked like to the farmer on the quad bike, waving in the gloom in my town-meets-country get up?

This is the divide – right there – when our romantic notions of a beautiful country walk crash headfirst into the reality of a working countryside. I can well understand why some urban visitors to hardcore agricultural areas might find it a little bit disappointing, and difficult to navigate and feel very short-changed on the beatific rural escape they pictured in their minds. This is blatantly not a tourist spot – it's a hard-working, intensively managed farmed landscape. No different, really, to going for a walk around an industrial estate – except this just happens to be in the countryside. A field of sheep and turnips is not set up for me and my dog to have a lovely time (even if there is a public right of way). Its primary purpose is to produce food. And I'm OK with that. I don't go there again. I buy an Ordnance Survey map, find a country park and some nice woodland, and we have a much nicer time there.

But some would argue the countryside is as much for me and my dog as it is for the farmer and his sheep. I'm on a public footpath, I'm being a responsible dog owner and I'm respecting the Countryside Code. It is my right to roam on these muddy turnips in the dark!

In May 2021 I caught an interview with the Right to Roam campaign on BBC Radio 4's *Farming Today*. My ears pricked up when I heard the words 'all' and 'full' in the context of public access. 'Do they want to open up the entire countryside?' I wondered. I had to find out.

According to the campaign's website, the 'majority of the English countryside is out of bounds for most of its population. 92% of the countryside and 97% of rivers are off limits to the public.'

This lack of access to green space, they believe, is having an extraordinarily damaging effect on the health of our nation and people should be encouraged, not criminalised, for wanting to reconnect with the countryside around them. They make an impassioned plea: 'We urgently need access to nature, its beauty, its space, its flora and its fauna, for our health, our creativity and our peace of mind. In a world of steel, glass and concrete, of stress, ecological detachment and screen-based lifestyles, the countryside is a natural health service that can heal us.'

The solution they are calling for is to expand the reach of the Countryside & Rights of Way (CRoW) Act in England, increasing the public's right to walk in woodlands, along the banks of rivers (swimming and paddling included), across 'all downland' and the Green Belt. The latter, due to its proximity to urban centres, is highlighted as 'possibly the most important' because it would help those who most struggle to connect with green spaces: 30 million people living in towns and cities.

I have lots of sympathy. It's frustrating trying to go for a hike in the area surrounding my parents' farm. In the early autumn of 2019, fresh from a week's hiking through the Snowdonia National Park, my sister Nicola and I were filled with rambling zeal. We just didn't want to stop so decided to plot a route around our home on the OS Explorer. We quickly regretted it. It wasn't like the Snowdonia Way, with all its tourist-friendly trail-markers and sturdy stiles. Footpath signs had weathered away or fallen down, we teetered precariously on wobbly, rotting stiles, gingerly negotiated brambles and barbed wire, and sometimes the path was blocked off completely – which felt suspiciously deliberate.

We got barked at by scary farm dogs and felt a little spooked by several 'KEEP OUT: PRIVATE' signs in the woods. It came as a huge shock, and some disappointment, to us that the area we know, and love, isn't the most welcoming place to walkers.

So, I can understand why the Right to Roam campaign has chosen such a bullish approach. The website does not tiptoe around:

'One thing about rambling is that it abhors a fence. For this reason, we will present a campaign than actively eschews partisan divisions.'

I confess this bit did make me laugh out loud. I picture myself walking over our neighbours' fields at home:

'Excuse me sir! I abhor this fence!'

Some of the farmers I know would turn purple – utterly apoplectic with rage. It would set the public access debate back a hundred years.

You don't have to be a genius to work out why. Aside from the basic human drive to guard one's own – 'This is my land!' – ramblers, dog walkers and visitors to the countryside haven't exactly helped matters. In short, some people (no, a lot of people) act like total dicks in the countryside – and it's ruining the experience for everyone else. If you've been unfortunate enough to be on the receiving end of an irate farmer or landowner, hollering at you for not sticking to the path or keeping your dog on a lead, you won't have been the first – and you won't be the last.

I'll never forget something a hill farmer in the Peak District said to me about exactly this: 'Every day I wake up and think, "I'm going to make a real effort to be nice to the visitors today", and I start off well. The first two or three walkers go past with their dogs off the lead, and I'll politely remind them, and maybe even stop for a friendly chat. By the third or fourth it's wearing a bit thin. By the ninth or tenth I'm at the end of my tether and usually say something I regret.'

Stephen Ware, who farms near the gorgeous village of Weobley in Herefordshire, was verbally abused and physically threatened three times in the space of one month, simply for asking people to stay on the footpath or keep their dogs on a lead around the lapwing nesting site, which is under a stewardship scheme:

'People have been locked away for so long during the pandemic and it's like some kind of psychosis. Dealing with them is an absolute nightmare,' he says. 'We are planning a PR campaign with the local estate in the hope it will help.'

The footpaths on George Young's farm in Essex got a lot busier in lockdown. In fact, the whole place became a free-for-all. He received a complaint from someone who twisted their ankle in a muddy rut on one of the private farm tracks. They moaned it was too bumpy to walk on properly, which is unsurprising really considering it's used for heavy farm machinery:

'It's not a public footpath!' says George in exasperation. 'And the ruts are there because we can no longer drive down the main road through the village because it's too busy, so we have to run all our machinery through the farm.'

So, he's taken his tractors off the road out of consideration for commuters and villagers, but still gets it in the neck about making ruts in his own land? The man has got to drive his tractors somewhere, people!

You could argue that George has more than a thousand acres to call his own and doesn't have a clue what it's like to spend lockdown in a pokey flat or a house without a garden. Our mid-terrace in Bristol only has a tiny backyard and the concrete smells of Lucy's wee. For us, getting out every day wasn't just about exercise: it was basic survival. Our mental and physical health depended on it. Green space was our sanctuary and millions of people, just like us, felt the same.

But George does get it, he does understand. In fact, he wanted to provide a lot more access than the existing three miles or so of public footpaths around his farm:

'In 2020 I was aiming for full "Right to Roam" on the farm with three simple rules: never enter a field with livestock in it; your dog must be on a lead at all times; and stick to the margins of the fields. My realisation since has been that you can't trust people. It doesn't matter what you tell them or what you give them, they will always take the piss. Now I've got the responsibility of my own herd, it has completely changed my attitude towards public access.'

'You don't want to give people the Right to Roam anymore?'

'No, which is a shame because I think it's important for people to have access, but I can't trust them.'

In my personal view, we as British dog walkers and ramblers are doing pretty well for ourselves. American farmers I've spoken to can't get their head around the idea of someone walking over their private land.

'I watch a lot of foreign TV on Netflix and watching stuff out of the UK is just crazy,' says Wade Dooley on a Zoom call from Iowa.

'They'll show whole groups of people wandering across a sheep pasture or something and it just blows my mind! Public access is a completely foreign concept to us. Property rights are sacrosanct. Back in the day people would get shot at for trespassing.'

So maybe we should be thankful for the 91,000 miles of footpaths and 20,000 miles of bridleways we have to run, walk, and ride across. Lucy and I have been discovering lots of new walks around Shrewsbury, including a permissive footpath along the river through several sheep fields allowing us to do a circular route back to the car park. The gates are well-oiled, light, and swingy and all the stiles are solid and easy to climb over. I say a conscious mental 'thank you' to the farmer or landowner every time – because they don't have to do this; it probably causes them a few headaches in fact. I keep Lucy on the lead out of respect and gratitude. I'm the guest here.

If we want more of our countryside opened up – I believe it's on us, the public, to show farmers and landowners they can trust us. Let's bring down the number of dog attacks first, reduce the littering and fly-tipping and demonstrate we can handle the responsibility we're asking for. Otherwise, why would you give a naughty child more treats? Because leaving your litter on a riverbank is childish behaviour, so is acting like an entitled jerk on someone else's private land or letting your dog off the lead around their livestock.

What if you're 100% sure you have your dog under control? I've had this debate many times with friends. Well, take Lucy for instance. I know she is scared of cattle and more interested in chasing sticks, so I trust her when they're at a distance – but if those cows come closer, she's straight back on the lead. Even though I trust her implicitly, I don't know the cattle. Farm animals can be unpredictable. With sheep, I wouldn't let her off the lead for a second. She's never shown much interest in them but if an animal runs, Lucy will chase it. It's hardwired into even the most compliant dog to chase things that run away from them, and you have no chance of stopping them once their blood's up in the chase.

If we want more access, then it's down to us – the general public – to prove ourselves, to show we can respect the land we want access to. The ball is in our court.

Equally, farmers and landowners must give the public the chance to prove themselves by keeping an open mind. By remembering the many thousands of people who cherish the land, and respect it, just as much as the person who owns it. The city children who write about their weekends in the countryside in school on Monday – excitedly telling their teachers about the animals they saw, the farmer they met, the fields they played in. The land you work and farm every day, to a family somewhere, is a holiday, a day out, a memory, a haven, a special place.

That's an amazingly powerful connection – and one to be cultivated and cherished in a world where farmer and consumer have grown so far apart. It's important to keep this in perspective: the majority of people who walk across Stephen's farm don't verbally abuse him. Most would shake his hand.

Let's assume we can heal all these rifts – that everyone, be they urban or rural, farmer or non-farmer, gets on together. It's well within reach – there's nothing stopping us overcoming our differences – but the success of a mass migration to the countryside will depend on something far more essential to modern life than good neighbourly relations: digital connectivity.

The 2020 Rightmove survey indicated that good internet and a spare room for a home office could overtake transport links on the list of buyers' requirements. Rightmove said: 'Fast internet has always been important but now it will be a must.'

Ah. That immediately knocks out huge swathes of rural Britain where fast internet is but a dream. I tear my hair out when I go home to the farm for a few days and try to work:

'Mum! The internet's down again!'

'It's the atmospherics!'

'What the hell are atmospherics?'

'I don't know but it's them!'

Perhaps feeling at the mercy of some invisible mystical force is preferable to dealing with EE.

In its Levelling Up Rural Britain report, published in February 2021, the National Farmers' Union lays it bare: poor connectivity has put rural areas at a disadvantage. When it surveyed its membership about digital technology, 'almost all respondents said access to reliable broadband and a mobile signal was essential for their business, yet less than half felt their mobile signal was sufficient for their business needs, while only 40% of farmers felt their broadband speeds were sufficient'.

There is absolutely no reason why every part of the UK shouldn't have fast internet. OK, so we have a few big hills, and some people live up bumpy tracks. Still no excuse. I would have had more sympathy with the 'remote and challenging landscapes' argument before I visited Olga Downs in the middle of the Australian outback, where I'm guessing there are more kangaroos than people and the nearest optician or clothes store is 400 miles away.

'In terms of population density there might be one person to every 50 square kilometres at a guess, and there are definitely more kangaroos than people,' smiles William Harrington, the fourth generation of his family to graze cattle on these 40,000 acres.

Olga Downs captures my heart from the moment I arrive. It's late when I finally step out of William's truck and place my feet on the dusty, amber earth of North West Queensland, after a 31-hour flight from the UK and a six-hour drive inland from Townsville Airport. The evening stillness warms my air-conditioned skin and frogs croak repetitively in the darkness – they sound comical, like children pretending to burp. I close my eyes, relishing another sound that spells adventure: insects singing in the night. When I look up, an unpolluted sky ablaze with stars overwhelms me with its massiveness.

I've been fascinated with the Australian bush for as long as I can remember, ever since I watched my favourite film of all time: *Crocodile Dundee*. For my thirtieth birthday I dressed up as Mick Dundee at a music festival in Devon. I sculpted crocodile teeth out of FIMO for my leather bush hat and carried a plastic hunting knife around. My friends all wore bright green crocodile onesies.

So, this is more than just a work trip for the BBC World Service, producing a programme about digital connectivity in remote regions. It's practically a pilgrimage.

Crocodile Dundee is based on a story of survival in extreme isolation. Mick, played by Paul Hogan, finds himself lost and wounded in the outback after a crocodile attack, spends weeks crawling around the bush, living on grubs and spit-roasted iguana, and miraculously finds his way back to Walkabout Creek, and a nice cold beer.

William Harrington may not have wrestled a croc, but he is also fighting to survive in extreme isolation. He's fighting for Australia's most rural communities, which could easily get left behind and forgotten in a rapidly advancing digital world. These days, without the internet, you're as lost as Mick in the woods:

'We want to keep people in the bush in Australia and we're not going to be able to do that without adequate infrastructure,' he says simply.

William, as well as working on the family cattle station, is an extraordinarily talented electronic engineer, with all the eccentricity of the genius inventor. He went to university in Townsville and is now doing a PhD, looking at how farmers use the internet in remote regions.

My visit coincides with the November mustering season, when the Harringtons gather their 2,000 head of cattle with the help of an aeroplane, helicopters, a cavalry of quad bikes on the ground, and me bringing up the rear in a battered old car. William roars past on his four-wheeler – I note he's the only member of the family wearing a helmet. The large white orb stands out against all the cowboy hats. It doesn't look as cool, but I immediately tut myself for thinking it. The UK's Health and Safety Executive is constantly on at our stubbornly cavalier farmers who thunder around on quad bikes without helmets. It tells me something about William's character – he isn't afraid to be different.

The herd are driven back to Olga Downs in a noisy early morning procession, a tumult of mooing and revving engines. I'm thankful for the racket as we near our destination; the dust outside the cattle yards kicks up so thick and blindingly, I can only navigate by sound, trusting the cattle know where they're going.

It's still only 8am – we've been up since first light to beat the heat – and the Harringtons are keen to push on and get the cattle drafted before the sun rises to its fiery zenith; it's only a few hours until this oven is preheated and ready to cook us all.

Today, we're pulling out the 'weaners', large calves ready to leave their mums, and the 'fats', plump, barren cows that have come to the end of their breeding life and will be sent for slaughter. The 'bush cows' and breeding bulls will be driven back up the station to continue grazing and mating and roaming around thousands of acres as if they were wild animals. The cows give birth alone and unassisted out in the bush, and the whole cycle begins again.

Later that afternoon, we retreat into the shade. I gratefully accept an ice-cold beer and sit with my feet up on the veranda, letting sleepy, dusty eyes drift hazily into a half-snooze. I snap myself awake. I'm a journalist not a jackeroo – I'm supposed to be working. I head off in search of William.

I find him in his air-conditioned workshop surrounded by dozens of computers and monitors, circuit boards and various bits of metal and wire, with a drill in his hand. Amidst all this tech, it tickles me greatly that the fire alarm keeps beeping and William hasn't thought to replace the battery. He's so engrossed in riveting brackets on to the back of flood monitoring cameras, he notices nothing else.

William's water surveillance system was his first electronic farm diversification – cameras mounted next to water troughs and rivers which send regular images of the water level to a phone or desktop, via an app. It allows cattle graziers to check on their troughs without having to get in their vehicles and physically drive around each one – which can take hours and even days on the biggest cattle stations. Local authorities use William's cameras to monitor creeks and rivers as an early flood warning system, allowing them to close the roads before it gets too dangerous. William's inventions potentially save lives, on top of enormous amounts of labour. His cameras cut carbon emissions from running vehicles and substantially lower costs for family businesses. It's a win/win.

However, the app only works if you have internet and William quickly discovered most cattle stations in his region only had patchy satellite coverage, or none at all. He was also frustrated with his own connection at Olga Downs. This is a guy who literally builds computers. Not being able to get online, for him, was a joke.

The big internet companies weren't going to break a sweat hooking up a spattering of remote cattle stations to superfast broadband – so

William took matters into his own hands. In 2016 he set up Wi-Sky – a wireless provider delivering fast internet to people living in the outback. How it works is amazing. Through intense negotiations and considerable expense, he managed to tap into a fibre broadband cable running through Richmond, the nearest small town. He then bounces the signal around 30 internet towers spread across more than 62,000 square miles of North West Queensland (an area larger than Scotland).

They're nothing fancy. The internet tower on the highest point of Olga Downs, the first one he ever built, is an old radio tower used by the Americans during World War Two. It used to stand on Castle Hill in Townsville, but William bought it 'for a good bottle of whisky', transported it home and gave it a new lease of life.

There's another one of his towers serving the one-horse town of McKinlay, better known as Walkabout Creek, where they filmed the pub scenes in Crocodile Dundee. It's a four-hour drive further inland from Olga Downs – a thrilling expedition through the kind of far-flung country I've always dreamed of. A flat expanse of sandy earth where the horizon, puckered only by a few thorny trees, meets a sky of the most intense, unbroken blue. We pass a rusty old, abandoned truck on the side of the road – it could have been there 50 years or more. There are trees growing out of it now. We drink a beer in the Walkabout Creek Hotel, take some photos, buy some souvenirs, and drive all the way back again. I have honestly never felt happier.

But the occasional *Crocodile Dundee* superfan isn't going to keep McKinlay's one and only tourist attraction afloat. The Walkabout Creek Hotel is a strange little place – half dedicated to its Dundee fame with sun-bleached photos of cast members on the wall and bits of the original set carefully preserved, like precious museum pieces. The other half is as functional and no frills as a Travelodge motel. Lots of miners stay there, passing through to work in the nearby silver, lead and zinc mines.

McKinlay, though beautiful to me, feels like a faded town, down on its luck. Perhaps the arrival of superfast internet will bring more prosperity to its residents than the fading memory of a 1980s film.

'Doing our bit to provide equitable access to the internet, to try and help keep people in these remote regions, has been incredibly rewarding. It sounds simple but it's something people out here have never been able to have before,' says William, describing the intense joy his customers experience when they can watch Netflix for the first time.

The internet has seeped into every part of our lives, so much so it has become a basic human right, central to twenty-first century existence – even the UN says so. If we continue to leave our most rural communities with poor or non-existent connectivity, it's effectively a proclamation to the world that they matter less than people in the city. From the moment I stood in the Australian outback, opened WhatsApp, and sent Mum a photo of me behind the bar in Walkabout Creek, I realised there really are no excuses. We need to be more William about Wi-fi.

He grins happily:

'In many cases ours is better than what you can get in a lot of the capital cities which is really good fun to be able to say.'

Just think about it. If Walkabout Creek – immortalised as one of the remotest corners of our planet – has superfast broadband, why can't Mum watch *Corrie* on catch-up, three miles outside Oswestry? It's not the atmospherics. It's a legacy of rural communities being overlooked, undervalued, and left behind.

By allowing a digital divide to open up in the first place, successive UK governments have created a much costlier and more complicated problem further down the line. How do we make a system that's biased towards urban populations now work for a minority group that were left out from the very start? We're flapping that stable door long after the horse has bolted.

A thought struck me recently: *Crocodile Dundee* is fundamentally a story of urban/rural divide. A New York journalist travels to the Australian outback and finds herself awkwardly out of place in a completely alien culture, trekking for hours on end through an inhospitable landscape in a thong swimming costume. She places her faith in a rough and ready bushman who mocks her city 'Sheila' ways, but also introduces her to the wonders of the natural world and saves her from a crocodile.

She returns the favour by inviting him to New York City and introduces him to the wonders of escalators, hot dogs, and pedestrian crossings. He strolls around the city being naïve and saying inappropriate things. Ultimately, though, it's a love story. The bushman falls for the city girl, they live happily ever after and the producers get two sequels out of the franchise (never, ever watch the third one).

The moral of the story is this: difference works. When these two worlds collide, beautiful things can happen.

There's a real-life Mick Dundee and Sue Charlton living just down the road from Mum and Dad, though I seriously doubt Debbie Jones walks around Rhydycroesau in a thong swimming costume.

She was more of a 'briefcase, stockings and heels' woman during her many years working for Barclays bank and, latterly, the NHS in Bedfordshire. In 2001, she'd had enough of offices and uncomfortable shoes. Debbie and her husband decided to leave Dunstable and start a new life in the countryside. In hindsight it was a valiant effort to save their unhappy marriage, but at the time, it felt like the adventure they both needed. They bought a house in the Candy Woods near the tiny village of Rhydycroesau on the Welsh Marches, the borderlands between England and Wales.

Debbie, a self-confessed 'larger than life person', also bought a Stagecoach franchise – a part-time performing arts school for children aged four to 18. A lifelong thesp, she'd always wanted to teach drama, and this was the perfect opportunity to reinvent herself. Word soon spread among the locals that a theatre professional had moved into the area and Debbie was duly roped into directing the Rhydycroesau Pantomime, an institution since 1978. It was a good way to make new friends, so off she went, hurrying down to the village hall on dark December nights to corral local farmers who hadn't learnt their lines.

Debbie threw herself into community life and even bought a few greyface Dartmoor sheep, but was more than happy to admit she didn't have a clue what she was doing.

'Ask Gwynfor. He'll help,' one of the locals down the pub suggested.

Gwynfor Jones had heard there were new people living in the Candy, from down south apparently. He didn't mind – newcomers bring new ideas. It's good for the area. He was an incomer of sorts himself, born six miles away in Llanarmon. He moved to Rhydycroesau in 1956, when he was four years old, after his parents bought his grandparents' tenanted farm from the big estate. Lord knows how many generations of his family came before them. It was built in the early eighteenth-century, with traditional wooden outbuildings including a hayloft over a narrow passageway – just wide enough for a horse and cart – leading into the yard and down to the farmhouse, where '1732' is carved in stone on the front wall.

Gwynfor and his dad milked cows until the 1970s – only about 20, which was far too small to make it worthwhile, so they knocked dairying on the head and concentrated on sheep and suckler cows instead. Gwynfor ran the farm alone when his mother passed away in 1992, followed by his father in 1997. He'd never married – never met the right person – and he'd been far too busy on the farm anyway to worry about finding a partner. He was a bit of a workaholic in those days. If love doesn't happen, it doesn't happen – no point dwelling on it.

The new woman from the Candy seems nice. He decided to help her with her sheep.

'There's a bubble coming out!' Debbie shrieked down the phone to Gwynfor, the quiet bachelor farmer from up the road.

'Yup. That's normal when they're lambing.'

'Oh, OK. Phew.'

Debbie and her husband weren't getting on very well. The move to the countryside was supposed to bring them closer together, but they just seemed to be drifting further apart. It had been five years – they'd given it a good go – maybe it would be better to sell the house in the Candy and go their separate ways. Gwynfor helped them both move. He loaded their divvied-up belongings into his trailer and drove his neighbours to their new homes in opposite directions. Debbie moved eight miles away.

A year passed. Gwynfor wasn't sure how long one should wait before asking a lady out after a divorce. He and Debbie had been friends for years now – but would it be appropriate to invite her to dinner? He was shy around women. He didn't talk much, and he'd been single his entire life. He needed some Dutch courage.

Debbie said yes.

'Oh bugger,' said Gwynfor.

They went to The Sun Inn for their first date, just outside Llangollen. It was a karaoke night, which Gwynfor definitely had not planned. Debbie found it funny. They made each other laugh a lot that evening. They knew straight away they had found something special together.

When Debbie moved in, she was met with a beautiful, hardworking farmhouse complete with a giant inglenook fireplace, a tall oak dresser loaded with Gwynfor's mother's willow pattern crockery, exposed beams, and original wonky windows from 1732.

They were a bit draughty, but Gwynfor didn't feel the cold. He didn't spend much time thinking about the house or its period features – it provided a roof over his head and that was about it. He'd nailed up some wallpaper to cover a damp patch in the lounge and the vintage orange bath worked perfectly well without being attached to the wall in the bathroom. There was only one plug socket in the entire upstairs.

'Where can I plug in my tongs, and hairdryer, and toothbrush charger, and bedside table lamps and all the things I have?' asked Debbie.

'We didn't have electric up here until 1962,' Gwynfor pointed out, thinking she should be thankful for the one socket.

When Debbie offered to do some laundry, he wheeled out an ancient Hoover single tub washing machine with an electric mangle on top. It took her all day.

'This isn't going to work Gwynfor,' she said, exhausted.

They decided to pool their resources and do the house up a bit. Debbie bought a new washing machine.

Gwynfor's marriage proposal was subtle and shy. It happened in 2014 while driving north on their annual holiday to Keswick in the Lake District. On passing a sign for Gretna Green, he turned to look at Debbie and said:

'Shall we go a bit further?'

That's as close as he got to: 'Will you marry me?'

Luckily, she knew exactly what he meant.

At four o'clock in the afternoon, to the incessant sound of Scottish bagpipes and with two South African tourists as witnesses, Debbie and Gwynfor said: 'I do' in Gretna Green.

They celebrated with champagne and ginger nut biscuits.

He was 62.

'I didn't meet her until late in life, but I like everything you do.'

Gwynfor looks over at Debbie from his armchair by the fireplace and smiles:

'You get the best out of everything.'

It's the longest sentence I've heard Gwynfor say – he's possibly the quietest man I've ever met. I wonder if he's more comfortable speaking in his native Welsh tongue, or whether it's just how he is. Even Debbie, who talks a lot, is momentarily speechless.

'Oh, Gwynfor . . .'

She's bashful, and delighted.

'He's my Henry V,' she adds with a flourish, wafting her glass of fizz around and bursting into Shakespeare. 'Act 5, Scene 2: A good leg will shrink, a straight back stoop, a black beard turn white, a curly head grow bald, an attractive face grow wrinkled and a pretty eye hollow. But a good heart, is the sun and the moon.'

I look over at Gwynfor.

'Have you read *Henry V*?'

'Nope. Just *Farmers' Weekly*.'

We're sitting in the lounge on big squashy red velvet sofas. This is where they sleep in shifts during the three busiest weeks of lambing, taking it in turns to check on the ewes through the night.

'We get a good fire going, don't we?' he says.

They lamb 250 ewes between them and, when it starts tailing off, Debbie goes back to sleep in the bedroom. She used to keep a walkie talkie next to the bed so Gwynfor could buzz through if he needed help in the lambing shed or out in the field, but they couldn't get used to saying 'roger' and 'over' without bursting into giggles, so came up with a more basic form of communication – a big stick:

'He'll tap on the bedroom window with the Knocking Stick if there's a problem, and I always get up don't I Gwynfor? I never complain.'

He simply smiles, neither confirming nor denying.

It's getting late – I'd better be off home. I thank them for the glass of 'Nosecco' and hand Debbie my glass.

'It's not bad, is it?' says Gwynfor, draining the last of his and getting to his feet. That's another new one on me – watching a Welsh hill farmer sipping bubbles in his slippers on a Thursday evening.

They walk me out to my car. It's still light and the birds are trilling merrily, making the most of the long summer days.

'Careful you don't bottom your car out on the track,' says Debbie. 'Just drive up the field if it's easier.'

They stand behind the little iron and chicken wire gate in front of their ivy-clad porch. A black and white farm cat slinks along the garden wall and curls up next to them, tail swishing lazily in the evening sun.

I open the door to my car and, just before I get in, a thought strikes me: 'Gwynfor, have you noticed that a lot of visitors, or people who move into the area, tend to be quite . . . talkative?'

He nods and smiles:

'We see them on the footpaths coming up from the Candy and we always have a chat with them when they come past.'

'They're chatty?'

'Oh yes!'

'Why?'

'Because they come from the town. They like it here. They find it interesting.'

'I think they want a bit of it,' adds Debbie. 'They want to engage with you, so they can feel part of it too.'

I bump along the drive in first gear, tufty grass tickling the bottom of my car, and think about the love story I've just heard between the 'Local' and the 'Incomer'. To me, there's only one thing that makes them obviously different: words.

Debbie says lots. Gwynfor says very few.

It dawns on me that this is one of the starkest and most obvious differences between my two worlds, yet I've never consciously acknowledged it.

Many a time, my sisters and I have watched in fascination the long, companionable silences between farmers in the livestock markets. They stand together in little groups, stoic and calm, hardly saying a word to each other. They seem to communicate in barely audible 'ayes' and 'ah wells'. It's enviable really. As we loiter awkwardly, finding their silence excruciating, they are content to just. . . be. They don't feel the need to waste words.

Urban people, or certainly many people that move to the countryside, just like Debbie, tend to talk more. A lot more. They engage freely and joyfully, with curiosity and warmth. They are open books – happy to put themselves out there, happy to be known, and they wish to know you in return.

'They ask a lot of questions,' says Dad.

'What do you mean?'

'Well, local people keep themselves to themselves. They don't go poking their noses into other people's business, but newcomers want to know why you do things the way you do – things I've never really thought about. They're interested in everything.'

And then he smiles.

'It's quite nice I suppose.'

Because, secretly, Dad loves a chat. And a chat is all it takes.

EPILOGUE

I RUN MY FINGERS ALONG THE SMOOTH, weathered old door into the granary at Craigllwyn Farm. There's a round hole, just below the latch, and I bend down to peer through it. A forgotten memory bursts forth, like a jack-in-the-box, flipping my tummy with its force: I remember standing here on my tiptoes as a child, giggling with my cousins, spying on Grandad Bill as he scooped feed into buckets for the animals. He knew we were there, stifling laughter behind the door, but pretended not to hear us so as not to ruin our game.

We'd soon get bored of peeping and off we'd run to climb on the bales in the barn, making slides in the straw. He'd tell us off if he caught us:

'Down from there you lot!'

I can see him now in his green overalls and wellies looking up at us from the floor of the barn, his kind face not used to shouting.

I lift the latch and step through the door. Hazy memories pull into focus. Two little girls, my cousin Hayley and I, huddled in a square pen of hay bales nursing a pet lamb we called Bright Eyes. The malty smell of the meal house. Tin water bowls hanging over concrete cattle stalls.

All ghosts. All history.

The outer wall of the granary has gone. It's open-sided now, leading into smart cattle sheds and modern steel buildings. The domed corrugated roof of the old barn is still there, but it's been added to, extended tenfold. Clean white concrete floors have replaced the weedy, stony yard I remember as a child.

'This was a potato field,' says Dad, wandering over to an enormous new machinery shed, like an aircraft hangar, housing rows of vintage tractors – mostly Massey Fergusons. We wander inside, being nosy.

'These are all collectors' items.'

Dad stops next to a Massey Ferguson 35; produced in the UK from

1956 to 1964. It's cute and tiny and red. The round little headlamps remind me of Brum from the children's TV series. He pats the bodywork:

'Now this is the tractor I used to love driving in the sixties. One just like this with the harvest trailer behind it. I drove it all the way down the Racecourse into Oswestry without touching the brakes.'

He looks around at the scale of everything and shakes his head:

'If Dad could come back and see this now . . . he'd be dumbstruck.'

'How do you feel?'

'It's the past now.'

We lean on the gate overlooking the fields towards Llansilin. There's a good view of The Hill from here. I look out towards my home: my heaf, my belonging. Tears swell, threatening to spill out. I don't want Dad to see. I feel silly.

'Are you coming?'

He's already walking away, down the yard, towards the farmhouse. I wipe my eyes and follow on, through the wicket gate, past the old pig sties and the garden with the walnut tree Dad and Grandad planted in 1960.

I've ran along this path a thousand times, thundering down the steps and through the old green door into Nan and Grandad's back kitchen.

'Slow down!' Nan would scold, turning round from the Aga with the kettle in her hand.

There's a posh extension on the kitchen now, and new oak doors. We knock. Dogs bark. My cousin Emma opens the door.

'Hiya! Come on in.'

'Cup of tea?' My auntie Julie pops the kettle on the Aga. Still the same one.

'Aye, please.'

Dad plonks himself at the end of kitchen table. My uncle Stephen walks in.

He waited two decades to buy back Craigllwyn Farm. The ambition lodged in his mind on that spring afternoon in 1991 when he and Dad reluctantly stopped bidding, unable to beat the £90,000 the land made at auction.

'I was determined to get it back if it came up for sale,' he says. 'It motivated me.'

He worked, and worked, and worked: not farming but fixing and selling tractors. I suspect it was the driving force behind his phenomenally successful agricultural engineering business: there isn't much my uncle doesn't know about Massey Ferguson tractors.

The house came up for sale in 2010, the land and buildings in 2014. Stephen put in offers that couldn't be refused as soon as he saw the For Sale signs.

'Joy, I suppose,' he says in his shy, reserved way when I ask him to describe the feeling.

'Just . . . joy.'

I watch these two brothers, chatting over mugs of tea in a room full of memories, where our family have shared meals, and talked, and cried, and laughed since the 1920s. Nanny and Grandad Bill smile down from a black and white photo on the wall.

It's going dark outside. I check my watch – I need to drive back to Shrewsbury.

'How are you settling into your new place?' Stephen asks.

'Yeah, I love it.'

'Nice town, Shrewsbury,' Dad nods sagely.

'Closer to home,' says Stephen, with a gentle smile in my direction. Yes. I am.

THE END

ACKNOWLEDGEMENTS

THIS IS A BOOK ABOUT PEOPLE. I could not have written it without their time, openness and the thoughts and feelings shared in more than 50 interviews. A heartfelt thank you goes to each and every person named, and unnamed. Due to Covid, many interviews happened over Zoom and some of you I have never met face-to-face, which I hope won't always be the case. I like to imagine bringing you all together in one place for a big party – and what a beautiful and diverse group we would be. Definitely no more divide!

Thank you to my publisher Joanna Copestick for spotting something in me and giving me a second chance. From one idea which didn't work out, came a book straight from the heart. Thank you Samhita Foria for your gentle guidance, and all the team at Kyle Books for showing a new author how it works.

This book has been in my mind for a long time, without me ever realising it. Only when I sat down to write, did the words and feelings come spilling out. So many of the experiences I've had, the places I've visited and the people I've met are thanks to my career at the BBC. Much of what I've learned about farming and the countryside has come from working on *Countryfile* and *Farming Today*. I have been lucky enough to work for fantastic, encouraging bosses who nurtured me from researcher to producer, and alongside brilliant colleagues. I have made friends for life and travelled the world. For all that I will be eternally grateful to the BBC.

I owe a great deal to the Nuffield Farming Scholarships Trust, which gave me a life-changing opportunity in 2016, along with my sponsors the Royal Welsh Agricultural Society and Trehane Trust. It provided a 'way back' and reconnected me with a community, and a part of myself, I felt in danger of losing. How much I have learned and developed since becoming a Nuffield Farming Scholar. Thank you.

To my wonderful friends – RPR in Brum, Pub Club in Bristol,

Ben and Borm, the Triumvirate of Terror and Sister Club – you are the people who have shaped me more than you know. So many of the thoughts and anecdotes in this book came out of our chats, laughs, debates and adventures. A special thank you to Emma Campbell and Ceri Davies for reading the early chapters and spurring me on to write the rest.

To my family past and present – cousins, aunties, uncles, grandparents, ancestors – I feel immensely proud of who we are and where we come from. Our shared roots and heritage provided the inspiration for this book. I wish I could tell Nan Beryl. I hope she knows.

To Mum and Dad – 'thank you' doesn't even begin to cover the gratitude I feel for growing up in such a safe and happy home, the unconditional love and support I've known all my life, and the strength and resilience you instilled in us. Mum – you are the reason I write. Thank you for sharing your love of words and stories and laughter. Dad – I could not have written this book without you. Subconsciously, I have been researching it since I was a little girl. Thank you for all the conversations we've shared on journeys in the pick-up, for always answering my questions and encouraging me to be curious about people, and the world.

And, lastly, to Alex. My patient city boy. You believe in me even when I don't believe in myself and keep me going when I feel like giving up – I don't know how you do it but thank you for doing it. Thank you for giving me the room to retreat into myself, and write, and for being there when I re-emerge. Your hugs are the best. I love you.

Anna Jones is a journalist, broadcaster, blogger and Nuffield Farming Scholar. She can be heard on BBC Radio 4's *Farming Today*, *On Your Farm*, *Costing the Earth* and BBC World Service. She worked on BBC One's *Countryfile* for more than a decade, and still occasionally produces programmes. She has written for the *Guardian* and the farming trade press. Growing up on the Welsh Borders, from at least five generations of farmers on her father's side and a long line of butchers and farm labourers on her mother's, Anna's heritage is deeply rooted in working class, conservative, rural values.